What Is Information?

ELECTRONIC MEDIATIONS

Series Editors: N. Katherine Hayles, Peter Krapp, Rita Raley, and Samuel Weber
Founding Editor: Mark Poster

(continued on page 189)

WHAT IS INFORMATION?

Peter Janich Translated by Eric Hayot and Lea Pao

ELECTRONIC MEDIATIONS 55

University of Minnesota Press
MINNEAPOLIS LONDON

Published by the University of Minnesota Press
111 Third Avenue South, Suite 290
Minneapolis, MN 55401-2520
http://www.upress.umn.edu

Printed in the United States of America on acid-free paper

The University of Minnesota is an equal-opportunity educator and employer.

22 21 20 19 18   10 9 8 7 6 5 4 3 2 1

Library of Congress Cataloging-in-Publication Data
Names: Janich, Peter, author. | Hayot, Eric, translator. | Pao, Lea, translator.
Title: What is information? / Peter Janich; translated by Eric Hayot and Lea Pao.
Other titles: Was ist Information? English
Description: Minneapolis: University of Minnesota Press, [2018] | Series: Electronic mediations; 55 | Includes bibliographical references and index.
Identifiers: LCCN 2017022054| ISBN 978-1-5179-0008-3 (hc) | ISBN 978-1-5179-0009-0 (pb)
Subjects: LCSH: Information science–Philosophy. | Communication–Philosophy.
Classification: LCC Z665 .J24 2018 | DDC 020.1–dc23
LC record available at https://lccn.loc.gov/2017022054

To the memory of Fernand Hayot: father, reader, scientist

CONTENTS

TRANSLATORS' INTRODUCTION

Eric Hayot and Lea Pao

Today, "information" is a concept with an extraordinarily broad reach. It touches on various aspects of engineering, where its lineage goes back to Claude Shannon's 1948 "Mathematical Theory of Communication." Via engineering, information enters the arenas of theoretical physics (where it is construed at times as the primary ontological unit of the universe), evolutionary biology (where it figures the relation between organism and environment), and cybernetics (whose feedback loops and recursive structures amble toward the dream of artificial intelligence). "Information" in a post-Shannonian sense also figures in sociology (in the work of Niklas Luhmann or Harold Garfinkel) and in various iterations of the concept of the "information society" (the term makes its first appearance in Masuda Yoneji's 1976 *The Information Society as Post-industrial Society*; by 2003, it was the theme for a United Nations–sponsored World Summit in Geneva). "Information" is a major theme in the field of library science (and the schools of information that foster it); it circulates around the edges of a wide variety of media theory; and it makes its appearance in contemporary anxieties about the rise of Big Data, the promise and threat of the information "cloud," and the rise of quantification as a mode of social dominance and control.[1]

Despite various attempts to produce a "unified theory of information" (Hofkirchner) or to define it as a single overarching philosophical concept (Capurro, Floridi), this cacophony can hardly be said to have constituted itself as a single field. We continue to live in an age of multiple, overlapping "informations," some of which don't talk to each other at all. There is—or perhaps we should say there "is," since the status of each of these forms of being is very much in play—the information in the public library; the information Google or Facebook has about you; the information classified and secreted by state actors; the information guarded by copyright and patent law; the information on the World Wide Web; the information exchanged between living beings and their environment; the information encoded and decoded in DNA, whose metaphorization as the book or grammar of life is complete; and the information stored and communicated by quantum states. There "is" also information as opposed to knowledge, as opposed to wisdom, as opposed to data, as opposed to experience, information more or less "raw" and more or less "processed." There are, in sum, informations, various interlocking concepts that come together in partially stable information regimes, each of which collocates a set of terms, a set of ideologies, and a set of goals that determine its articulation of the concept and its attendant words.

Both in academia and in the popular imagination, our sense of this new life with and as information, this intensification of informational life, seems to stem from a series of recent technological transformations—most prominently the rise of digital methods of information encoding, compression, storage, and networking. But "information" can also be understood as a matter of human concern for far longer, dating back to what Michael E. Hobart and Zachary Schiffman call the earlier "information ages" characterized by the development of writing (in Mesopotamia, 3500 B.C.E.) or the European development of the printing press, or even further back than that, if one follows, instead, the thinking in Claude Lévi-Strauss's *The Savage Mind* (this latter influenced by its author's contact with Shannon and postwar engineering-based information theory, as Bernard D. Geoghegan has shown).[2] As Alfred Lord and Milman Parry demonstrated in their analyses of the written works of Homer, the oral poets who recited those lines in preliterate Greece did not do so entirely from memory but built their individual performances from algorithmic structures that allowed them to act flexibly and creatively within a larger system.[3] We know, too, that the patterns of drumming used for long-distance communication in Africa employed structures of redundancy designed to compensate for the relatively small throughput of their communication technology.[4] And we know from the earliest written discussions of the value of writing that literacy was considered to have potentially salutary and potentially damaging social, political, and personal effects, ranging from the stunting of the individual capacity for memory to the mimicry of authoritative knowledge. Hence Socrates' worrying, to Phaedrus, that writing might diminish the mind's ability to hold things clearly before it; hence the great book-burning project of the Chinese emperor Qin Shi Huang (260–210 B.C.E.); hence indeed the whole history of attempts to regulate, control, and expand the various processes whereby humans store, organize, or transmit ideas about the world to one another, or about one another to the world.

None of this longer history obviates the interest of the contemporary information-moment, of recognizing both the potentially transformative action of the new technologies and the import of the tremendous rise in political and philosophical discourse about them. But we do so better when we do not imagine that "information" has only just arrived onto the world-historical scene. Informing practices have long been with us. The mistake is to think that they were invented only in the 1960s, or to imagine that only now, at this moment in human history, have the various social activities and concepts associated with information—problems of storage and memory, of language and communication, of authority and canonicity, of medium and genre, of epistemology and ontology, of singularity and generality, of the categorization of the world or the organization of those categories; of the forms of reference and meaning attributable to quantity over quality, of the capacity for transmitted communication to be animated by life itself—become relevant to the great humanistic project of understanding the species, its societies, and their histories.

The Foucauldian epistemic transformation or the Kuhnian paradigm shift that would divide us, like so many denizens of Babel, from an innate comprehension of our past has, to be sure, a certain anthropological charm. The premoderns were so unlike us, we imagine, that they lived in a different world. And yet, we think we understand them well enough, even if we do not always grant them the same capacity to understand us. Epistemic breaks, historical revolutions, the rise of new ages: all these are simply a matter of a frame of reference. Changing one's perspective is enough to turn any difference of kind into a difference of degree. The seemingly vast gulfs between Bretons and Normans, Swabians and Hessians, collapse easily enough into holistic national identities during the Olympics or the World Cup. The chasms dividing the moderns from the premoderns, the calligraphers from the interpreters of Morse code from the word processors, can likewise be made to disappear if one enlarges the scope of analysis, reaching, at the limit, the vast comparable framework of the life of human beings on Earth. The twin confines of the species and the planet are, after all, "historical" in the strict sense of the word. They emerge in time and are fundamentally *of* it. From their perspective, simple patterns of mutual compatibility and social practice appear that are not—from a certain perspective—fundamentally altered by the arrival of this or that new technology. The decision about what changes and what stays the same, about the degree to which history "breaks" at any given moment between a definitive after and a receding before, is not a matter of history itself (which bears no traces of such breaks). Breaks are rather always *for* something; the similarities and differences they invest in their actors, events, and timescales serve an insistent and almost certainly immediate purpose.

Lately, at least, some of the forces most interested in declaring an absolute break with the past—Everything is new! Everything is information!—believe, for political reasons that are all too comprehensible, that any such break will put paid not only to the human sciences but to the university more generally. For who needs the humanities, who indeed needs any of the vast realms of knowledge and information produced by the species in the tens of thousands of years since its emergence, if every kind of knowledge that matters has been produced in the past thirty years? Who needs to read Newton or Darwin or Smith (not to mention Aristotle or the *Ramayana* or the letters of Catherine the Great) if what they thought or said or did has been surpassed by the things we know now? Who needs the past at all, and indeed who needs the present, if what we learn tomorrow will once again render its past obsolete?

Such positions are not themselves the pure products of a love for historical novelty; they have a variety of motivations. We must nonetheless observe that the declaration of absolute historical newness— we are living in an "information society," in an "information age"— sustains and undergirds a casual stupidity and carelessness. The fantasy of a rededication of the social to economic necessity and of the idea of learning to preparation for a lifetime of paid work, though it does not need to operate hand in hand with a dismissal of those aspects of the university devoted to the whole of human life or to the acquisition of "pure" knowledge, nonetheless does so consistently enough. The successes of this violence are being enacted today on the lives and futures of the universities and their employees as well as on the current and future students whose freedom and capacity for change depend on a rich education.

It is therefore for both political reasons and because we believe that it says things that are true that we bring you this little book. Its author is Peter Janich, a German philosopher of science. Janich (1942–2016) studied physics, philosophy, and psychology at the universities of Erlangen and Hamburg, earning a PhD in philosophy in 1969. After some time teaching at the University of Konstanz, Janich served from 1980 to his mandatory retirement in 2007 as professor of systematic philosophy at the University of Marburg. The author of some twenty books and more than two hundred articles, Janich continued to write after his retirement, publishing seven new books in the last decade of his life. Though his work has been translated into Italian, Korean, Japanese, and Chinese, he remains little known outside the German context.

Janich began his career as a member of the Erlangen school of philosophy of science. Drawing on the work of the mathematician and logician Paul Lorenzen (1915–1994) and of the historian, philosopher, and musicologist Wilhelm Kamlah (1905–1976), Erlanger constructivism *(Erlanger Konstruktivismus),* or methodical constructivism *(methodischer Konstruktivismus),* as it was later known, intervened in the philosophy of science by rejecting both scientific naturalism, the idea that the work of science describes facts of nature that occupy some reality outside that of human life, and cultural relativism, the notion that all truth-claims depend entirely on the social context within which they emerge. Lorenzen and Kamlah aimed initially (in *Logische Propädeutik: Vorschule des vernünftigen Redens,* 1967) to reconstruct the philosophical ground of science, and especially of logic and geometry. This meant moving away from strong epistemological claims about the nature of reality, because those ground their claims in the unprovable givenness of reality to the mind, and from systems grounded in axiomatic claims, which, in blocking the justification of the system at the axiom, generate merely hypothetical formalisms. Lorenzen and Kamlah favored instead an ordered, stepwise reconstruction of basic concepts grounded in practices of human interaction. This involved developing logic on the basis of systems of rules guiding two participants in a conversation—in seeing logic, that is, not as a matter of idealized or purely formalized practice but as emerging from, and philosophically groundable in, actual human behavior. In this way, members of the Erlangen school hoped to produce noncircular explanations for the basic concepts that logicians, mathematicians, and scientists use to do their work.

It was in this context that Janich did his earliest writing, publishing books laying out the proto-scientific grounds for concepts of time (in *Die Protophysik der Zeit,* 1969) and space (*Euklids Erbe: Ist der Raum drei-dimensional?,* 1989) and, later, matter (*Das Maß der Dinge: Protophysik von Raum, Zeit und Materie,* 1997). In all these cases, his concern was to emphasize the degree to which science, and the instruments of science, expresses human intentions and must be understood as the product of interests, and the expression or fulfillment of those interests, by sequences of actions. As Janich puts it in the preface to the second edition of *Protophysics of Time,* "every bit of knowledge of nature discovered through experiments, i.e. through instruments, is created by the success or failure of human intentions which were realized by the instruments."[5] The proto-physical work of establishing a foundation for the practice of physics does not therefore assert propositions but describes norms that allow for the expression of social intentions and for their materialization in instruments designed to achieve those intentions. Proto-physics, Janich writes, "asserts nothing about nature, but rather about production processes as suitable means for particular ends."[6]

This at least explains the "constructivist" part of methodical constructivism. As for "methodical," it refers to an absolute insistence on the *sequencing of actions* in practice as a feature of the action as such, and thus to the idea that this sequential order constitutes the action. To give an example Janich borrows from Hugo Dingler (1881–1954), if one wants a painted statue, one does not first paint the statue then carve the wood. The order (1) carve wood, (2) paint carved wood is a matter of method, and a methodical understanding of the statue as such must include it *in that order* if it is to come to grips with the actual statue as it emerges in historical time and human practice and not to mistakenly describe a metaphysical version of the statue "as such." For Janich, the statue "as such" includes, in other words, the ordered process of its making; the limit of the idealization of the statue lies in an understanding of its production in steps by time-bound, goal-pursuing human beings. Because all practices belonging to the ordinary lifeworld of human beings happen, like the making of the statue, in some kind of order, the actual existence of these practices in historical fact allows their "reconstruction," that is, their methodical working through. And this in turn allows us to reconstruct adequate proto-figures of scientific concepts and techniques, all the while resisting the temptation to consider the finished concept as the "origin" or "truth" of the practice that led to it. (One important side effect of Janich's approach is that it undoes any strong distinction between modernity and premodernity or, for that matter, between civilizations; the scale of the human community, and the vision of human history as continuous and striated rather than structured by breaks, produces a strong emphasis on continuity rather than rupture.)

In the mid-1990s, Janich began to use the term *methodical culturalism (methodischer Kulturalismus)*, distinguishing his work and that of his students at the University of Marburg from the constructivism of the Erlangen group (which had by then spread across Germany). The methodical culturalist program is laid out in a collection of essays that Janich edited with his former student Dirk Hartmann in 1996. Describing the shift as a "substantial, serious *expansion of the goals and means of philosophical work*," Janich and Hartmann distance themselves, in their Introduction, from Lorenzen's "ortholinguistic" emphasis on the development of logically sanitized terms for the production of logic, geometry, proto-physics, and the like.[7] A too-narrow focus on the rectification of names, they argue, misses out on the fact that language always operates relative to the goals it seeks to accomplish. Language is thus embedded in cultural practices, an action like any other and thus a matter of pragmatic, goal-oriented behavior that takes places in specific historical contexts. While methodical culturalism retains the earlier emphasis on basing its principles on "complete, non-circular chains of arguments," as Janich writes elsewhere, it embeds "this business of foundation and justification in communities which share forms of a prediscursive consensus by historical processes of (cultural) selection."[8] The culturalist turn pushes Janich's work further in the direction of considering the historical facticity of human practices in ordinary daily life as a fundamental component of the fundamental as such. It also brings him quite close, in its emphasis on practices, methodical order, and communities of consensus, to the recently rediscovered work on information by the American sociologist Harold Garfinkel.[9]

Janich's resistance to metaphysics in either its philosophically idealist or scientifically naturalist guise thus leads to his insistence on the embeddedness of human thinking and human activity in sociocultural contexts that must, he argues, be included in any attempt to understand their ideational (conceptual) or material (technological) consequences. This comes close, both in *What Is Information?* and in his other work, to a rough materialist ontology of the social, even if the word "ontology" never appears in his writing. Indeed, Janich's work is profoundly Kantian in its refusal to consider the existence of anything outside its conceptualization by specific beings in specific historical contexts. It also resists, though usually with far less vigor and engagement with sources, what it refers to as "postmodern" relativism (often emblematized by the work of Richard Rorty).

If we think, nonetheless, of ontology as a kind of limit on conceptualization, a place where the work of thinking might both begin and come to rest—that is, not as an idealized or purely conceptual essence but as a ground for thought—then one might decide to see in Janich's work something like a "social" ontology. Such an ontology would be historical in three respects: first, in the sense of being located in a specific physicotemporal situation that includes both the passage of time and a variety of social mechanisms for relating to it; second, in relation to the human embeddedness within the general physical structures of both the planet and the universe; and third, in its inclusion of the social structures through which humans work to achieve goals in those contexts. This line between "essence" and "foundation" is a fine one to walk, as Janich himself has noted: his goal is "not to take refuge in various forms of relativisms," he writes, but it is also "not to claim too much in the form of absolute, last or eternal foundations."[10] But one can find its interlocutors, as in, for instance, Dipesh Chakrabarty's observation that the fact that no human society we know of has absolutely excluded religion means that any theory of the social that does the same may not in fact be theorizing the social "as such."[11] What would it mean to make the historical facticity of the human lifeworld a ground for humanist thought? Janich offers one sustained attempt to work out such a program.

As this summary of Janich's career and thought makes clear, his writing takes place almost entirely within German-language philosophy of science. His major intellectual touchstones lie exclusively in the German philosophical tradition leading up to the work of Immanuel Kant; in the twentieth century, the figures with whom he most engages, besides other contemporary German philosophers of science, belong largely to the pre– and post–Vienna Circle group of thinkers who launched the fields of analytic philosophy and philosophy of science in both the German and the Anglo-American academy (Rudolf Carnap, Carl Hempel, and others) and to contemporary theorists and historians of science writing mainly in German and English. For readers attuned to the fields in which "information" is most intensely under intellectual discussion in the United States—which include library and information science but also and perhaps especially those literary and cultural zones influenced by media studies—this intellectual inheritance will be, as it was for the translators of this book, profoundly unfamiliar. Janich, who once said in conversation that "the continental philosophers think I'm an analytic philosopher, and the analytic ones think I'm continental," begins from a background that does not include any of the philosophy done under the rubric of theory, whether it be French (Foucault, Derrida, Lyotard), German (Benjamin, Adorno, Sloterdijk), or English (Butler, Spivak, Williams). He similarly makes little or no reference in this book or elsewhere to work in media theory in any language (no Kittler, no Flusser), nor does he put himself in conversation with the tradition of sociological studies of science (Latour, Rabinow).

This tangential relation to American academic humanism makes this book both interesting and hard to read. On one hand, its unfamiliarity produced, in our own first reading of the work, moments of incomprehension, or even dismissal, as certain intellectual investments or concerns taking place within the frame of Janich's own thought simply failed to touch us in any place that mattered. On the other, we experienced moments of intense theoretical excitement, of radical proximity, moments in which Janich's way of thinking seemed either to echo or to solve problems that have been before us in our own fields.

Two places where the encounter between Janich and United States–based literary and cultural studies seems most subject to a productive, overlapping distortion, a kind of uncanny overlap or *fuseau horaire,* are the question of language, and particularly of what one might think of as nonperformative or noncommunicative language, and the question of human–animal relations. When, late in the book, Janich excludes from consideration "the style of chatter or loquaciousness that belongs to the world of shallow entertainment" to focus on "morally serious communication that shapes the situation of living beings," he seems to run the risk of violating his own culturalist proscription. That the philosophical consideration of the use of language should proceed on the backs of only certain types of language may end up relying on a too-narrow vision of the social itself.[12] It was surely not the case, for instance, that in the earliest human societies, no one engaged in gossip or told dumb jokes, that these emerged only once the basic problem of survival had been solved, that it was only once everyone had enough to eat that *Homo sapiens* allowed members of its tribes to finally sit down and write a song or two. Everything we know suggests instead that the "unserious" forms of language (so often associated with women's talk or children's play) have been part of the social from the beginning.

Such forms of language (and perhaps the realm of the aesthetic more generally) are epistemologically useful, perhaps more epistemologically useful than the "normal" or "serious" cases of human activity, because they are a good source of outliers of all types. The aesthetic, after all, characterizes a set of practices that aim for (among other things) a deliberate *intensification* of experience as well as (in certain eras) an equally deliberate *experimentation* with it. Though they may seem only to supplement the normal case, then, aesthetic acts can also define the limits of social or conceptual possibility and hence establish the reach of the imagination (which has been, after all, also in the social from the beginning). Here it is not a question of finding a fatal flaw in Janich's program so much as opening the door to the inclusion of such practices, which have as solid a social ground as their moral, serious cousins, within the framework of action-schemas that Janich describes elsewhere in his work. What would a theory derived from an insistence on fundamental human rational orientation toward goals *(Zweckrationalität)* look like—what would the idea of a fundamental orientation toward goals look like—if it had to include among those goals the social practices of laughter, nonsense, artifice, or play?

As for the human and the animal, Janich mentions in this book a position he has laid out far more strongly elsewhere, of an absolutely strict differentiation between the human and other animals and a complete rejection of what he calls "overstretched biological theory" that likens the products of human culture to birds' nests or spider webs. Nonhuman animals, Janich argues, are not responsible for the consequences of their actions, nor are their actions "guided by cultural standards" that differ historically, geographically, socially, or politically across place and time.[13] Janich's position develops largely in response to work in evolutionary biology and evolutionary psychology that aims to reduce human culture to nature, that seeks, in other words, to make a scientifically naturalist, universalizing case for the identity of humans and animals.[14] Those working in literary and humanistic fields in the United States will be familiar with the critiques, mounted from the field of animal studies, of both a Janichian strict differentiation between human and nonhuman animals and of the kind of biologistic reduction he criticizes.[15] The question therefore is how Janich's own thought might fare were it to accept a more fluid boundary between, for instance, rational and nonrational action (as might be required, also, to allow for the presence of the elderly or the disabled in the realm of the human) or whether a defense of a stricter distinction in light of these possible critiques would make any difference to his work.[16] More broadly, the entire question of rationality encounters in both jokes and gossip, as well as in the questions of reflex versus reflection, automaticity versus consideration, that characterize a conventional disposition of the human–animal divide, a potential limit to its role as the fundamental ground for human social life. Human communities have, after all, always lived together and in antagonism with nonhuman animals as well as with jokes, puns, and the like. What would it take to develop a theory of culturalist rationality grounded in the historical facticity of that actual human experience?

Janich himself has noted that the expansion toward culturalism offers the philosophy of science an opportunity to tackle the realm of the aesthetic, which it has so far avoided for obvious reasons. This interest also borrows from another consistent theme in his work, a critique of the marginalization of labor and craftwork in philosophical writing. For Janich, this exclusion dates back to the Greeks, who assigned laborers (*banausoi* in Greek) to lower levels of the body politic (in Plato's *Republic*) or located craftwork at the lowest level of the ladder of human activity (in Aristotle's *Nicomachean Ethics*). The Greek word for this kind of work survives in the modern German word *Banause,* which refers to a "person with inadequate, shallow, narrow-minded views on intellectual or aesthetic matters; a person without an understanding of art, or with an unrefined style of life"—in other words, to a philistine or a peasant.[17] In keeping with his general interest in seeing the work of modern science as *Handwerk,* that is, as an extension of craft practices dating back to the earliest human communities, Janich begins therefore to refer to the work of philosophy (and intellectual life more generally) as *Mundwerk,* mouth work, attempting thereby to undermine the philosophical preference for theory over praxis, ideas over action. (This emphasis on action and practice oriented toward socially defined goals of care, communication, and activity may recall for some readers the vast artifactual elaborations of Elaine Scarry's *The Body in Pain.*) Janich's program is laid out at greatest length in his recent *Handwerk und Mundwerk* (2015), whose subtitle, *Über das Herstellen von Wissen (On the Production of Knowledge),* suggests the range of its ambition.

As for Janich's specific take on information, it follows. It is, as you will see, fairly relentlessly focused on the legacy of the postwar invention of the information-concept in the context of communications engineering, though it touches more generally on the ways in which that concept emerged hand in hand with ideas in semiotics, evolutionary biology, and emergence theory. All these, Janich argues, participate in the wholesale "naturalization" of the sciences, of human beings, and of information that he aims to destabilize here. In its place he aims to leave us with a theory of information as, first, a matter of social practice extending from the dawn of human culture to the present and, second, as a fairly narrowly defined, medium- and context-independent form of social exchange. Information, that is, is the "content" of an expression that is not altered, or ought not to be altered, when the contextual factors governing its transmission change. So, for instance, the "information" content of the bus schedule is the same whether one sings it or signs it, whether one is in a good mood or a bad one, whether one speaks in English or French, and so on. More crucially, as far as Janich is concerned, it does not make sense to speak of "information" without first establishing what one means by "communication." "Information" is something communicated; it is (in some cases) the "content" of communication. To understand why a society would need "information," then, requires us to understand why someone might want to *inform* someone else of something, to see what social function or socially determined needs are met by the development of an action-schema that determines the meaning and value of a particular action of informing.

This is a long way from the extensive metaphorizations of information in biology or physics. For Janich, there is no "information" in the shape of a fish fin, because the latter cannot be responsible for the act of communication that would inform us of it. The idea of information that emerges from this work thus draws a line around one major dimension of information mania. At the same time, Janich's attempt to ground the concept in human practice opens the door toward uses of information in a science that would no longer be defined by an implicit naturalism. In this way the book gives us a glimpse of one of the major goals of all of Janich's work, which is to reject the strongly institutionalized division between the sciences and the humanities and, indeed, to ground those opposed behemoths in the same philosophical space.

A word, finally, about the translation. This is a translation by a certain type of scholar for a certain audience, in a certain place and time. It does not reproduce the work; it alters it. In this way, this book is not (necessarily) a lesser version of Peter Janich's *Was ist Information?*, containing almost, but not quite, the same amount of information and ideas as you would find in the German. Nor is it (necessarily) a greater version, in which the movement into another language creates new echoes, new figurations, that match and then exceed the capacities of the original. The language of philosophy and the work of translation are a lesson, always, in the impossibilities of perfect informing. The translators are, in that respect, in no more control over every specific alteration (or exact reproduction) of this translation than Janich was over every specific meaning in the original, or than you could be, *felix lector*, in the reading of it.

Acknowledgments
The translators are grateful to Sam Frederick, Daniel Purdy, and Pia Pao, who indulgently answered various questions about points of diction and grammar, and to Annemarie and Peter Janich for their hospitality and support.

ＯＮＥ
INFORMATION AND MYTH

This book focuses on the history of a scientific myth that has for a long time now left the sciences behind and become part of daily life. As with many myths, this one entails effects whose causes are misunderstood; it affects ordinary views and opinions and takes part in the basic professional credos of scientific and philosophical fields. Its object is information. Operating as either a fundamental concept in theories of culture or as a description of the content of the genome, as the eponym for a young scientific field or as a political problem, as a benchmark for powerful new technology or as the ideal of a message of scientific salvation, the mythology of information has many forms, many narrators, and many addressees.

Information about Myths

Myths are, according to a general lexical sense, sacred stories. Literally, myth *(Legende)* refers to the text to be read.[1] Where a lexicon defines it, *Legenden* are described as "stories that do not correspond to certain truth."[2] Another reference book tells us that legends were a custom of the medieval Church, where they were read aloud on saints' days.[3] Such stories, part folk tale, part poetic workmanship, aimed to

1

manifest wonder, to animate the saint's life as example and ideal. The same thing happens in the mythology of *information as a natural object.* No matter what the encylopedias say, however, uncertain truths have never kept anyone from erecting places of pilgrimage. In those auratic locales, whether officially recognized or merely tolerated by the church, faith replaces doubt, and the self-appointed custodians of the saints tell of their many wonders. Neither the freedom of mythology's folkish mode of presentation nor the liberties of its poetic design have ever sullied fervent invocations of the holy.

Myths about Information

The myth discussed here comes together in one simple declaration of faith: "information is a natural object." No myth worth its weight in salt would consent, however, to exist in such an abbreviated form. Besides, what is a "natural object"? Everyone knows and uses the words *natural* and *object,* but most of us don't bother asking what they really mean or how we use them. The people who like asking such questions are mostly philosophers. But they are very much at odds with one another about what nature (or "the natural") is. Even the question of whether we should talk about nature with or without a definitive article—in German, the difference between *Natur* and *die Natur*—separates those who believe that there is a single nature, independent from its conceptualization by human beings, of which humanity forms a part, from the camp that thinks that nature should only be talked about as an adjective, hence *natural.* And *natural* is of course a word with many, many meanings, as one gathers from this incongruous list of its potential opposites: *artificial, technical, artistic, spiritual, civilized, affected, contrived, unhealthy,* and so on.

Before getting to a stripped-down retelling of the myth of information as natural object, before we see how information plays in that mythology a role almost from the big bang forward—including acting, in the primordial soup of an early, inhospitable Earth, as an origin for life itself—it would be good to get some solid ground under our feet. Information is a central concept not only for the empirical (natural) sciences but also for the technical sciences focused on information technology, which include computer and information science as well as sciences of communication and media, all of which emphasize information's mathematical and structural aspects. I will leave aside here the fact that in an information society we also talk about political and psychological aspects of information hegemony, about information as an economic good or a factor of production, about the survival value of the information gap, and so on.

There exist any number of books that treat the question "what is information?" within the broadest possible framework, moving from the classical history of the concept to the invention of the information sciences and the conceptual adoption of information in fields like physics, psychology, biology, cybernetics, economics, and sociology, and thence to the development of the information society and the political and ethical questions it implies.[4]

Between the extremes of encyclopedic comprehensiveness and the narrowness of a focus on a single topic lies the question of the origin of the myth of information as a natural object, the countless popular narratives that support it, and their widespread consequences. A decision between exhaustive coverage and a dangerously singular focus does not impose itself when, however, we are dealing with a clearly identifiable *intellectual and social failure,* as we are here in this book.

Failures have to take place somewhere. They must set themselves up, be caused by someone, attract some onlookers, and disturb some of them. Were one to give the state of affairs that results from this failure a simple name, it would be *the naturalization of information.* Its specific site is the philosophy of science, in particular, the philosophy of mathematics and of the natural and technical sciences. But it extends its reach to include all realms of professional academic philosophy as well as the uncountable armies of those who love philosophy, or think about it regularly, whether they are in academia or in journalism.

For an observer, a failure is what disturbs. The disturbance determines who defines the failure, who describes and analyzes it, and who, in the happiest of cases, resolves it.

The name of our failure, *the naturalization of information,* stands for a more general *program.* The argument that nature is exclusively or primarily an object of the natural sciences has been made programmatically. Now, using the word *naturalization* would make no sense if one wanted simply to assign something that is already natural to nature. *Naturalization* refers here to the claim made by the natural sciences to the full (and finally exclusive) responsibility for the study and mastery of information. In short, naturalization aims to make information only (or at least primarily) an object of the natural sciences. This aim is, naturally, long accomplished. To understand how and why, we need to take a closer look at the different narratives that make up the information myth.

When myths are hagiographies, they become in some way the verbal form of a pictorial representation, of *icons*. Myths need icons, just as complex narratives need their popular versions. One icon of the information myth is the famous phrase uttered by the father of cybernetics, Norbert Wiener, according to whom "information is information, not matter or energy."[5]

"Information is information" is certainly not the most successful form in which a science might express its principles. But the addition "not matter or energy" shows us what the sentence really means: the concept of information belongs to the same field of study as the concepts of matter or energy, and so ultimately to the field of physics. And the continuation of this famous citation, which argues that "no materialism which does not admit this [that information is information] can survive at the present day," makes the matter plain. It does Wiener's short cyberneticist credo no violence to interpret it as follows: information is not simply matter or energy but something autonomous, something with its own structure, a structure, one might add, that only appears via a process or an effect, just as the information impressed onto a vinyl record only appears when the record is played. Thinking this way, one would then say that the oscillations in acoustic pressure that take place when the record is played express as a structure of energy the spatial structure of the record, considered as a piece of matter. And so we see how information appears in its first iconic form, as something to be carried aloft like the plaster figure of a saint in an Easter parade. (It remains to be seen whether Wiener is speaking here about actual scientific objects or only about scientific concepts.)

Another icon of the modern mythology of information is *genetic information*. Ever since biological genetics and chemistry combined to create a new understanding of the molecular processes that take place in the cell nucleus when an ovum and sperm combine—ever since, that is, the second half of the twentieth century—it has become common practice to speak of these processes (and their influential models like the double helix, the gene, and their transformations) in the language of communications technology. The public announcement of the "unlocking" of the human genome in 2000 (by the geneticist Craig Venter and U.S. president Bill Clinton), itself the result of a combination of scientific, political, and commercial interests, produced a proliferation of representational metaphors in media reports—the book of life, the grammar of life, the writing of life, the script of life, and so on. But these figures of speech were not just for journalists; today, words like *coding, transcribing, translating, reading, speaking,* and so on, have become indispensable to the very representation of genetics. Their use is not a matter of the media finding figurative ways of explaining genetics to nonscientists but rather a reflection of the original language of the experts and of textbooks in the field. The lexicon of genetics, in other words, does not result from the development of simple illustrations of complex concepts for the unwashed masses. No, it reflects the *original scientific language* used at the frontiers of research; it formulates the account of the causal mechanisms whereby a living being acquires new characteristics from its parents in an idiom that has usually only been applied to speaking and writing human beings.

To this we must add a third icon and narrative of the myth of information. Its words are in common use. They belong to the general realm of epistolary correspondence and, for that reason, include centuries-old concepts like that of the "sender" or "receiver"; in Germany, thanks to the national postal service, they have become part of a certain statist officialese. Other words, such as *signal* or *code,* came into common use after the events of the Second World War, where they were not already known by those who worked on railways or had anything to do with the civilian transmission of the Morse alphabet. You don't need the slightest contact with theories of technical communication to use this vocabulary, which originates mainly in the practice of everyday life. The English word *code,* for instance, comes from the Latin *codex* (tree trunk, writing tablet, book), which appears in legal as well as ecclesiastical contexts.

Our third icon in the myth of naturalized information is therefore *communications technology.* The storage of census data on punch cards (using automated counting systems developed by Herman Hollerith), the invention of broadcasting, the telephone, and the teletype: all these marked the historical beginning that leads to our contemporary technological landscape, with its many forms of technical communication (understood only by experts) and dominated by the widespread use

of computers. Information's status as a natural object has been legiti-
mated by the astonishing success of two hundred years of engineering; what is technology if it is not the omnipresent and omnipotent
application of the laws of nature? (How did that lexicon put it? "Stories
that do not correspond to certain truth!" What if it's not a certain truth
that technology succeeds by applying the laws of nature? See chapters
2 and 5.)

A fourth, somewhat younger icon for our myth comes from *neuroscience.* The brain, information-processing machine and control center
for human and animal organisms, is the virtual poster child of natural
objects. Scientists would be all too happy to simulate its capabilities,
or to copy or translate them, as they attempt to do in research on artificial intelligence. Information-based models of the brain appear not
only where complex electrophysical or molecular–chemical processes
in its cellular compartments need to be simplified for nonscientists but
also wherever experts produce functional descriptions of the organ;
such models could not be built without the telecommunications-based
concept of information. When it comes to the supercomplexity of the
brain, the feeling of awe in the face of the sacred takes the form of an
astonishment before nature. The latter thus appears to us once again
as that sublime primary object, an object that humankind, with the
simple and pathetic linguistic means at its disposal, with its clumsy
technical models, can only ever imperfectly emulate.

With this final example the mythology of information as a natural
object acquires its complete contours. Its structure motivates any number of popular narratives. Information lies in *material structures of living
beings and machines.* We cannot understand that fact statically. We are
dealing neither with dead organisms nor with broken-down machines
but with those things in which information emerges, is transformed or
produced, sent or received, to form the basis for everything that takes
place in the higher spheres of human communication, from individual or supraindividual forms of consciousness to the sciences and the
arts, including poetry, music, and painting, indeed, the whole realm of
human culture. Contemporary technologies of culture alter our public
and private lives, transform economic and power relations, and create
new pictures of the world and humanity. They are merely the distant
tremors of an all-encompassing earthquake, in whose depths the material bearers of information find their ultimate foundation.

Mythological Critique: Information about Information
SOME PRELIMINARY REMARKS ON CRITICISM

The word *critique* does not, in ordinary German usage, sound very good. In the arts pages of the newspaper a literary critic may occasionally encounter an outburst like "Beat him to death! He's a critic!" "To criticize something" is generally understood to mean "to reject something." The critic's is the voice of condescension and arrogance, even when he can do the thing he criticizes no better than the person he critiques.

It's worth remembering, then, that the etymological origin of the modern word, the Greek *krinein,* meant originally to judge or to decide. In philosophy this original, ancient Greek meaning has inspired any number of concepts of critique, like the well-known three critiques of Immanuel Kant, which in no way reject anything, considering rather the natures of pure reason, practical reason, and judgment. It is in this Kantian sense that I intend to critique the information myth: the philosopher takes up the classic task of philosophy and practices a *critique in principle.* "In principle" means here—thinking critically now—merely to think things from their beginnings. My critique therefore has to do with judgments and decisions that lie at the origin of theories, and the historical and contemporary practices that stem from them. The criteria whereby we determine what a beginning is— whether the origin is temporal, definitional, logical, or methodical— will, of course, have to be named and justified as they come up.

Before critiquing the common ways of handling information as they are practiced in everyday speech and ordinary life, I want to reaffirm my earlier philosophical claims as they apply to the mathematical, natural– and technical–scientific treatment of information. Unlike other disciplines, in which the critique of concepts, theories, and methods of the goals of knowledge, and of the risks of applying it, form part of the daily business of researchers' work, the so-called exact sciences have historically been neither especially friendly to critique nor especially fond of it.

It would be an exaggeration to say that they include no critique at all. Any number of forward steps have taken place because a stronger argument has replaced a weaker claim or because of the substitution of erroneous assumptions by correct ones. But critique itself is neither emphasized nor rewarded. In mathematics, as in the natural and technical sciences, an institutionally established pattern of critique is not taken as a generally valuable part of the intellectual process. Scientists work in the mainstream. Someone who critiques what most of the people in a discipline take as given is usually suspected of not having properly understood things. The normal reaction to critique is therefore to proceed to elementary instruction in the basic principles of whatever concept is believed to be under attack.

This collective resistance to critique has a reasonable, or at least understandable, basis. It happens because critique does not direct itself toward the objective claims (the "results") made by these fields but toward their philosophical foundations, their modes of self-understanding, and the sociopsychologically interpretable statements that allow any given natural scientist or mathematician to understand herself as belonging to a specific discipline. A critique like the one in this book, that is, addresses not so much a set of scientific results, to which one could respond by referring to established methods. Instead, it risks feeling like an attack on the personal philosophy of the scientists themselves. Accordingly, in what follows, I want always to *strictly distinguish* between the critique of mathematical or scientific statements and a critique of their philosophical, psychological, or sociological side effects. (Regrettably, some philosophers who do not correctly distinguish between scientific results and the philosophies that ground them also come in for criticism.) Another way to say this is to say that we must always know what is being criticized: the object language of science or the metalinguistic philosophy (of scientists).[6] Following this brief detour through the concept of critique, and with an awareness of the risk that the critique that follows may provoke, in the sciences I address here, a certain emotional defensiveness, we are now ready to move on to the concept of information.

INFORMATION IN EVERYDAY SPEECH

We begin with a fairly overwhelming topic: the role the concept of information plays in ordinary life. The word is common enough to appear in all areas of daily routine. At the same time, it's worth noticing that this *use of information in ordinary speech is inconsistent and contradictory,* not only in German but in any of the other languages of modern Europe.

Anyone who obtains information about train schedules is obviously interested both in understanding the information they contain and in being assured of its reliability. In this sense, *information has meaning and value* for a person who seeks it or gives it out.

When two friends, on the other hand, talk about whether good old vinyl records or digital CDs make a better recording medium, they discuss stored or produced information in a way that attributes to it no inherent (word-)meaning or (sentence-)value. Instead, they focus on the preservation of signals in the recording or playing of the sound medium and on the assessment of loudspeakers and of the listening experience.

In other common uses of the term, like in the discourse on genetic information, no one will say, for instance, that someone with a genetic disorder like Down syndrome is "falsely informed," in the way that one might say the passenger who reads the wrong train schedule is.

And in certain situations in daily life we come across limit cases, in which we cannot tell whether we are dealing with information that has meaning and value (as it commonly does in linguistic statements) or information independent from those characteristics. A customer who buys a long-distance phone card that's good for a certain number of minutes expects the card to store information about how many minutes he has left, in the same way that a person uses pen and paper to balance a checkbook. It does not matter at all whether this expectation that the card stores information correctly can be expressed linguistically or whether the system "functions" like the gradual emptying of a bottle of water (or like anything else).

It seems clear, therefore, that *everyday language* is bound up with *two different traditions*: one the tradition of *linguistic communication* (which requires an attention to the functionality of information), the other the tradition of information technology focused on processes of transforming, coding, and decoding. These telecommunicative processes have a structure, but that structure does not have to be linguistic, meaningful, or capable of being true or false.

ON THE AGE OF CONTRADICTION

The myths and icons of naturalized information we have seen so far might seem to suggest that the newly germinating disciplines of the twentieth century—cybernetics or the mathematical theory of information, or their borrowings in biology, chemistry, and physics—are fundamentally responsible for this interesting ambiguity in ordinary speech. But that is not the case, or so a *glance at the history of concepts* tells us.

Already in the Latin etymology of the word *information* we find a certain doubleness. On one hand, *informare* refers to the crafting of a physical form (as in Virgil's description of the blacksmith's shield in book 8 of the *Aeneid*); at the same time, the word indicates the content of a complete sentence (as in Cicero). And if you leave the Latin roots of the words *information* and *to inform* behind, you can dig even deeper into the history of philosophy, arriving at the Greeks. In Aristotelian philosophy, *form* plays a double role as one of the three principles of the explanation of being, referring to the abstract form in which conceptual thought takes place, while also playing the role of a "cause of the form," in the sense of the spatial structure of a piece of marble that has been shaped by a sculptor.

None of this is inherently a problem—neither the fact of a word having two different meanings nor that a word has both literal and metaphorical uses. As it does today and in every age, context allows one to distinguish the differences. If we say someone "drew a blank," we don't mean that she used a pencil to trace an empty space—unless she's in an art class. The context determines what the words *drew* and *blank* mean, whether they refer to losing out in a lottery (their original, literal meaning) or to not being able to think of something (their common metaphorical use) or to the art class (a transposed literalism). In the case of these words we're dealing with the same word (with the same etymology) that has acquired two (or more) meanings. In cases where we see a homonym reflecting two different etymological trajectories (for instance, *well*, meaning a spring of water, comes from the same root as the modern German word *Wellen*, "waves," whereas *well*, as in "doing well," comes from the same root as the modern German *wohl*, with a similar meaning), the context determines, as in the first case, which meaning is in use. There is similarly no problem when, as in the case of metaphors, a *tertium comparationis* intervenes, introducing a third criterion of comparison.

What does cause a problem is the *equation of literal and metaphorical meanings* of "information" (as material or spiritual forming), such as it might be put into practice in (where else?) a classic text, in this case one by the inaugural figure of all modern philosophy, René Descartes.

In his "Arguments Proving the Existence of God and the Distinction between the Soul and the Body," one finds the view that in the human interpretation of meaning the forming of the mind takes place through an impression made upon the brain of a spatial form.[7] In this conceptualization lies the origin of a *mind–body problem* that continues to play an important role in the naturalization of information.

Ordinary German speech is saturated by the Cartesian bisection of the mental and the material. In the material world of the *res extensa,* actions are causally produced through pressure and collisions. When we say that something "makes an impression" *(Eindruck)* on us, or "imprints" *(einprägt)* itself on our minds, when "stimuli" *(Reize,* from *ritzen,* "to scratch"; the English may share a root with *stylus)* produce reactions, following the Newtonian model of action and reaction governing the collision between two billiard balls, we are using an image taken from the mechanistic displacement *(Verdrängung)* of physical bodies, even if this use is less metaphorical than Freud's, who speaks metaphorically of the mechanisms of repression (also *Verdrängung).* In carrying on the work of the scholastics, Descartes became the patron saint of the reduction of the mental to the material, especially when it came to the scientific explanation of the processes of human cognition.

THE DEVELOPMENT OF MODERN SCIENCE

In the three centuries that elapsed between Descartes and Claude Shannon and Warren Weaver's 1948 "mathematical theory of communication," the natural sciences took a series of enormous steps forward.[8] These steps began in classical physics and led to the development of a mechanistic program for the natural sciences more generally. One effect of this transformation was the integration of the study of human physiology, and of the human senses, into physics.

The success of the scientific point of view, a feat unparalleled in the history of human culture, had by the nineteenth century, however, led it into a comprehensive intellectual crisis, in which concepts and laws previously believed to be axiomatic and fundamental gradually came into question. Deep-seated orientations collapsed: mechanical and electrodynamic laws pointed to different, incompatible properties of transformation; statistical descriptions undermined deterministic ones; the division between organic and inorganic chemistry fell apart;

the origin and development of life in the field of evolutionary biology were subordinated to a schema of casual explanations; and two thousand years of certainty about the most certain of all theories, that of Euclid's geometry, wobbled under the influence of new non-Euclidean approaches.

Depending on your perspective, then, describing the nineteenth century as a century of science may highlight its successes, lauding the unstoppable triumph of empirical learning, or focus on the problems of its unsettled foundations. What would be decisive for the naturalization of information is that, at this time, the awareness of the problem in the disciplines of mathematics and natural sciences did not benefit from any contact with academic philosophy. Whereas one could still think of the philosophy of Immanuel Kant as a complement to Newton's physics, in the nineteenth century the natural philosophies of Fichte, Schelling, and Hegel constituted for researchers in the sciences mere occasions to establish boundaries between the two fields.

The tremors caused by the crises in the foundations of mathematics and logic, as well as those that resulted from the disruptions to the queen discipline, physics, by the arrival of relativity theory and quantum physics in the early twentieth century, brought a new kind of philosophy into play. Known today as the *philosophy of science,* it arose initially from reflections by experts in mathematics and the sciences into the linguistic and logical forms of the newly revolutionized, mathematical natural sciences. In the 1920s, the Vienna Circle's impressive attempt to build a unified science by means of logical empiricism is seen as a powerful revision that committed all scientific activity to following the principles of physics. Its first steps involved the rectification of language: the surgical amputation, via Occam's razor, of all superfluous metaphysical content, starting with the language of physics itself. Anything that could not, either by logical definition or experimentally verified causal relations, be explained through that language would no longer have a place in the bright sky of the scientific future.

For information, the consequences were (in brief) immense: the history of a literal and metaphorical discourse about the relation between *form* and *content,* scarcely two thousand years old, became, in the seventeenth century, a program for the explanation of content (perception, thought) by form (the material medium). The enthusiasm of the nineteenth century's successes in the empirical sciences meant that the research program governing this reduction would take place in the language of physics and in the mathematical theories it entails. Traditional concepts and theories of the mental or the cultural would, in such a program, be supplanted by the language of the sciences and by their fundamental treatment of the natural as the material.

Unsuspecting speakers of everyday German who sometimes understand information as having meaning and value, and sometimes as having nothing of the sort, are victims of philosophies that have leeched into ordinary language. *These philosophies themselves,* and not just the philosophical carelessness of communication scientists or cyberneticists, lie at the root of the naturalization of information. Together with the principles that ground them, they have established an intellectual legacy that emerges in the new forms of understanding of the natural sciences, in new conceptions of their mathematical language, and in new mechanizations of the idea of communication. The story of how that legacy came to be is told in the following chapter.

TWO
LEGACIES

The naturalization of information, I want to show, is neither a historically nor systematically isolated program or occurrence. It depends on related developments in the natural sciences, mathematics, and technology, each of these constituting a troublesome practical and conceptual inheritance. I will address them here under the rubrics of the naturalization of the natural sciences, the formalization of theory, and the mechanization of communication.

Before we begin, however, a note on the nearly instinctive manner in which basic shifts in the self-understanding of the sciences take place at all. No one in a field of knowledge like mathematics, mechanics, or electronics (to pick three areas important to the history of information technology) can understand the field, or move it forward, by strictly limiting himself to sentences describing the *objects* of that discipline. (By "objects," I mean things like plus or minus signs, or operations with those signs, in math; bodies, motion, or forces in mechanics; currents or programs in electronics. Object-sentences are about these objects, and so are scientific results, in their narrowest sense.) Even the simplest schoolbook requires, to make those concepts clear, a wide variety metalinguistic concepts (and sentences containing them), things like "principle," "axiom," "definition," "evidence," "theorem," "measurement," "experiment," "system," "performance," and so on.

If you look at the way they talk about themselves and their objects, schools of thought in arithmetic, mechanics, or electronics seem, therefore, to require *at least two distinct types of sentences,* which can be thought of as sentences about disciplinary objects and sentences about disciplinary methods. But no science that wishes to be studied or practiced can exhaust itself fully in these sentence types. Anyone who wants to understand scientific results and the metalinguistic concepts that govern them *must also engage with the philosophical terrain* in which any particular science is embedded. If you do not want to feel outraged that mathematics today requires formal–axiomatic, ungrounded first principles, and all of the undefined expressions and sentence structures that go along with them, then you have to swallow along with any discussion of contemporary mathematics the particular forms of philosophical knowledge that underlie, authorize, and legitimize them. In other words, you have to become a *formalist.*

Likewise, if you don't want to feel outraged that all your questions about important results in physics seem to be answered in advance by the assumption that the best possible results in the field result from a laborious process of empirical testing and observation, you have to buy in to the special forms of philosophical knowledge that ground physics. This means becoming an *empiricist.*

And (third) if you don't want to feel frustrated when you are confronted by the conceptual circularity of a theory of electrical networks whose understanding of the functions of electrical and electronic hardware dictates its understanding of the logic of circuits, even as it grounds its understanding of the logic of circuits in its understanding of the functions of electronic hardware, well, you'll need a philosophy appropriate to that belief system. So you'll have to believe in *systems theory.* In short, no mathematical, technical, or scientific knowledge can be understood, taught, and circulated unless its most basic doctrines are grounded in a philosophy appropriate to it.

These relationships are more problematic in the so-called exact sciences, since the philosophies without which they remain essentially unthinkable seem no longer to require any justification and are therefore almost never mentioned at all. Their unspoken acceptance functions as what a social psychologist would call a scent mark, a kind of unconscious odor that marks one's membership in a discipline or intellectual community. Joining the guild of mathematicians, physicists, or engineers requires knowing certain theories, and acknowledging certain results, but also speaking the language of whatever counts as mainstream philosophy for that group. This subordination is a necessary part of every scientist's life; it produces not only mechanisms of interpersonal socialization but also the framing and bracketing of allowable (and unallowable) questions.

The reader of the first chapter may already have wondered what justifies its disrespectful attitude toward informational "myths" and where in fact all this criticism of the basic premises of natural science and their relation to information is going. At this point I can propose a first answer: the philosophical critique of information this book develops does not want (or need) to argue that science produces invalid results. *Not a single scientific sentence or theory can be called false (or true) without reference to the particular philosophical forms that ground it.* It is those philosophical forms that have been most important to the misunderstanding of information as a concept, that are, accordingly, the major subject of this book.

Who Naturalized the Natural Sciences?

That the natural sciences required naturalizing seems like a paradox: the very name of the concept "natural sciences" suggests that these sciences have nature as their object. What can it mean to say that they have been "naturalized"? Earlier I defined the naturalization of information as the process whereby information became primarily or exclusively the object of the natural sciences. So the naturalization of the natural sciences must refer to the process whereby those sciences themselves were made into their own object. Let me explain how that happened.

If you date the beginning of modern physics from the astronomy of Johannes Kepler and the mechanics of Galileo Galilei and Isaac Newton—each one marking a significant departure from the principles of Aristotelian natural philosophy that dominated Europe through the seventeenth century—then you can assign the naturalization of the natural sciences to the very first moments of modern scientific thought. The rise of mechanical observation and measurement (in the work of Kepler or Tycho Brahe) and of real-world experimentation (by

Galileo, among others) constitutes a major sign of this transformation. Already there we can see the first steps toward a philosophical catastrophe: Galileo mistakes the operations and functions of his experimental machinery for the actions, and laws, of nature itself. Whereas Aristotle had distinguished (in a manner still perfectly comprehensible today) the natural from the artificial (*technē*, in Greek), with the latter emerging through tactical, purposeful human action and the former bearing on its own the philosophical responsibility for essences and changes, modern physics imagined that the physical action of its instruments—governed, after all, by the same laws of nature that they were discovering—pointed the way to a reconciliation of esthesis and the natural. It thus mistook its empirical and experimental results for properties of nature itself.

This misunderstanding represents a philosophical catastrophe because it overlooks essential conditions for the production of scientific knowledge, effectively exiling those conditions from the entire (self-)understanding of the natural sciences. Although no researcher of the last four hundred years would deny that experiments, measurements, and observations can only produce results when the usual instruments work "properly" (that is, as they are "supposed" to), discussing the actual *function and purposefulness of the entire research apparatus* has nonetheless become entirely taboo.

The history of naturalization thus begins from *seeing the instruments as natural facts* and believing that the function of those instruments stems exclusively from the causal determinations produced in them by the laws of nature. From this point of view, it's easy to overlook the fact that broken or miscalibrated instruments, ones that produce no useful experimental results, *point to underlying natural laws* just as much as their useful, properly calibrated counterparts. The active "natural laws" that seem to be at work in the means of scientific knowledge (that is, the instruments) cannot define the difference between broken and working instrumentation, which means that they cannot explain, either, how the knowledge they (seem to produce) has been gained. A simple example: a broken clock produces no useful measurement of time, even though it operates under the same "natural laws" (e.g., the oscillation of a quartz) as a working clock. (You can tell it does, because, among other reasons, those same laws explain the clock's brokenness and can be used to repair it.)

This mistake is, by my account, the classical error of all naturalization. From its perspective we learn that physics cannot generate its
own epistemology. Its various means of knowledge production, the
indispensable technical instruments of measurement and observation, the various mechanisms at the heart of physical experiments, are
taken inevitably, and mistakenly, as natural facts. And those natural
facts, physics tells us, can in turn only be understood and tested by the
instruments themselves. What the instruments' brokenness reveals is
a breach or gap in the intellectual project that instruments in general
propose. The experimenter (not to mention the designer and fabrica-
tor of the instrument itself) falls into that breach, and into error.

The history of naturalization produced, after the seventeenth century, a significant leap forward in the self-understanding of the discipline of physics. To understand it, we need to look at the work of
the physicist Heinrich Hertz. The preface to his 1894 *The Principles of
Mechanics,* often cited by both philosophers and physicists, lays out a
series of hypotheses on the understanding of scientific evidence that
are still shared by the vast majority of physicists today.

Hertz, who was the first to introduce the concept of the "model"
to the natural sciences, took the study of nature to be an empirical, experimental project whose goal was to create "inner images or symbols
of external objects" and to do so "in such a manner that the necessary
consequents of the images in thought are always the images of the
necessary consequents in nature of the things pictured."[1] The *pictorial
representation of natural relationships through logical relationships,* Hertz
believed, *must* be possible. In this conceptualization of nature lurks the
second major step of our contemporary (though still mostly unrecognized) naturalization of the physical world. Hertz continued: "If these
propositions are to be absolutely fulfilled, there must be a certain conformity between nature and the human mind."[2]

"Naturalization" (built, in this second phase, around a specifically
structured theory of knowledge) means, since Hertz, that *humankind
can know nature, because nature has made the human mind its correspondent.* An idea that would only much later, in so-called evolutionary theory of knowledge developed by Konrad Lorenz or Gerhard Vollmer,
be taken as a subject for scientific investigation—namely, the idea that
the evolutionary process has shaped the human mind's capacity to

understand the natural world—thus appears here in Hertz's work as an incandescent omen.[3]

On what grounds could a reasonable physicist make such claims about the human mind? What would justify them? Hertz's answer: "Experience teaches us that the [pictorial–logical] requirement may be satisfied, and hence that such a conformity [between nature and the human mind] does in fact exist."[4]

What escapes Hertz, apparently, is that this type of "experience" is *historical,* or (best-case scenario) a matter of *humanistic inquiry,* but it is not *scientific.* The experiences physicists have when they measure and experiment cannot simultaneously be, epistemologically speaking, measures and experiments of the correspondences between measures and experiments and the human spirit. The scientific snake cannot eat its own tail. Such historical experiences (that is, as official "successes" of physics as a discipline) would have to be defined, explained, and justified (but which criteria define "success" in physics? how would you select them?) if they are not to become merely assertions of a disciplinary dogma.

The naturalization of physics in Hertz thus *makes nature the cause of its own knowability.* This produces a shift away from the earlier, Galilean naturalization, which took for granted the naturalness of scientific instruments, to a naturalization that depends on the naturalness of the human mind—and thus of the subjectivity of the scientific researcher himself. This becomes the ground of valid research in physics. According to Hertz, the *scientific method* is empirical; the object of physics becomes a reality that depends in no way on human knowledge or intervention. Reality exists in the form of "external objects" that generate "necessary consequents in nature," all these *epistemologically validated* in scientific discourse by the presumption of a physical correspondence between mind and reality. In Hertz, (1) *realism* takes over the role of determining the object of scientific inquiry, (2) *empiricism* takes over the role of choosing a scientific method, and (3) *naturalism* takes over the role of justifying the validity of the natural sciences. The sources of the naturalization of our modern information concepts are to be found here, in the workings of this epistemological inheritance.

Hertz's three-part theory gains much of its traction from the fact that it assumes that the presence of images of nature in the mind *pre-*

cedes their appearance in language. He distinguishes "theory" from the "representation of theory." Theory, he says, is something mental, intellectual. Only its "representations" are communicable, appearing in mathematical formulas, in words, in sentences, and in typesetting or recording systems. Only there do the economy and purpose of choosing concepts find their place, where they serve as the site of judgment for "the value of the representation of theories" of physics (and not for "the value of theories" as such). The mental images themselves are right or wrong, and the theories follow suit—they are whatever "can be decided by a clear 'yes' or 'no,'" Hertz says, "though only from the perspective of our present experience, themselves subject to revision on the basis of later, more mature experience."

And so we get the legacy of the naturalization of information: following the anti-Aristotelian misunderstanding of mechanics in the seventeenth century, whose avatar was Galileo, physics became naturalized as a kind of epistemological picture-theory, in which the correspondence between nature and the mind became the fundamental criterion of truth. By contrast, the linguistic formulations of physical knowledge remain secondary. Hertz thus leaves us with a *model of scientific epistemology that has three levels,* hierarchically built on something real, something mental, and something linguistic, that is, the true, the mental, and the communicable. *This triptych would be the trap into which twentieth-century information concepts would fall.*

Who Formalized Theory?

The nineteenth century bore witness to a stunning series of major interactions between individual scientific disciplines, including (in the field of mathematics) the discovery of set theory by Georg Cantor and the development of non-Euclidean geometries (Carl Friedrich Gauss, Nikolai Ivanovich Lobachevsky, János Bolyai) and their empirical interpretations (Bernd Riemann, Hermann von Helmholtz). Basic geometrical axioms and concepts, previously taken largely for granted, came under serious discussion, and that discussion had implications for theories in physics. The challenges posed by these new modes of thought could not be disputed by arguments taking place fully within the framework of mathematics or physics itself; they were of a fundamentally philosophical nature.

Gottlob Frege's invention of a language for symbolic logic launched the history of modern logic, in whose context Frege raised the question of the meaning of basic mathematical concepts like numbers or operations. The philosophical work of Ludwig Wittgenstein (whose *Tractatus Logico-Philosophicus* refers explicitly to Hertz's picture-theory!) and Bertrand Russell (in his lectures on "logical atomism") developed finally the intellectual program that would launch the work in logical empiricism made famous later on by the Vienna Circle.[5] This work determined the philosophical conceptualization of twentieth-century mathematics and physics. (The deviations of the "critical rationalism" of Karl Popper, which opposed a deductive method relying on falsifiability to the inductive one of the logical empiricists, are marginal here.)

The decisive impulse for the reconceptualization of *scientific discourse* in the twentieth century came from David Hilbert, who may well have been the most important mathematician of his time. Hilbert laid out a programmatic case arguing that mathematics could not free itself from the philosophical quarrels and basic epistemological conundrums of the nineteenth century unless it reconceived itself as a form of *structural knowledge*. Only the logical and formal elements of axioms and theorems, and not the semantic or empirical content of mathematical expressions, Hilbert argued, were the subject matter of mathematical science. The field should only recognize one criterion of mathematical validity: logical consistency. On these grounds the philosophy of mathematics, understood now solely as metamathematical or proof theory, should and could therefore only legitimately appear in a purely mathematical, purely formal language.

In this way Hilbert liberated—or at least appeared to liberate—the disciplines of logic and mathematics from the uneasy grasp of philosophy and from the difficult and fundamental questions the latter asked. "Appeared to liberate," because Hilbert's solution could not answer (and would never be able to answer) other important questions: about the definitions of basic geometrical concepts (Hugo Dingler), about the scope of the tools of proof theory (Kurt Gödel), or about how to decide between competing models of logic (Paul Lorenzen). But liberated nonetheless, since these questions failed to interrupt the larger process of the formalization of both mathematics and logic, and formalist mathematicians submitted, as part of their integration into a disciplinary community, to the new, scholarly character of the formal sciences. *Formalism thus also becomes part of the more general scientific inheritance* that includes naturalism, empiricism, and realism (in Hertz's sense). Formalism doesn't understand itself this way, of course; it sets itself forth, confidently and dogmatically, as a pure (and ahistorical) system, as a programmatic limitation of the vast field of mathematics and logic to the realm of the formal.

The myth of the naturalization of information pivots on this formalist turn. After Hilbert, questions about the meaning of the *content* of theoretical concepts (their roughly "semantic" function), about their relation to reality, no longer count as scientific. What's worse, the quasi-historical development of mathematics governed by this formal–axiomatic program led gradually to the popular belief that any kind of explicit determination of the basic concepts of (an exact, axiom-driven) theory was in principle impossible. Axioms would be (syntactically, that is, laid out in sentence form) the *first* sentences, "defined implicitly" by their own internal content and terms (these latter essentially signs pointing to basic concepts). The appearance of these terms in the axiom itself, and the grammatical relationships between them, are enough to narrow and determine their function and meaning. The determination of the concepts that *precede* the axioms—which the prominent theories of Euclid and Newton at least attempted to define—modern mathematics treats as either unnecessary or unrealizable; equally impossible is, naturally, any kind of *epistemological justification of axioms* (axioms that in other procedures function as logical derivatives from axioms themselves!). In contemporary science and in theories of science the problems I am posing here remain almost completely unknown or unrecognized; in their place a vulgarized version of Hans Albert's "Münchhausen trilemma," which abandons the question of truth entirely, stands guard against the complexities of genuine thought.[6]

At this point it's worth noting that Shannon and Weaver's communication theory of information presents itself as "mathematical" and that Norbert Wiener, the father of cybernetics, was a mathematician. We will have to discuss the combination of formalism and empiricism in both fields. Especially for the historical development of geometry, the separation between formal (mathematical) and empirical (physics-based) approaches had important doxographical effects. This division forms one of the major pillars of twentieth-century scientific epistemology. Albert Einstein recognizes it in his oft-cited lectures on geometry and experience: "as far as the propositions of mathematics refer to reality, they are not certain; and as far as they are certain, they do not refer to reality."[7] This admission of Einstein's also constitutes a core issue for the philosophy of logical empiricism, as its leading light, Rudolf Carnap, noted in 1956: "Mathematical geometry is *a priori*. Physical geometry is synthetic. No geometry is both. Indeed, if empiricism is accepted, there is no knowledge of any sort that is both *a priori* and synthetic."[8]

The separation of these two possible modes of knowledge in the history of geometry applies, mutatis mutandis, to any form of scientific knowledge, and thus also to any theory of information.

Who Mechanized Communication?

This third legacy takes place, at least initially, within the general field of technological development rather than in the sciences or philosophy; it is, for that reason, a bit more obscure. It has to do with the mechanization of the spoken word in human communication.

Let's contrast the generally received understanding of *mechanics* from a more literal sense derived from the Greek. Literally, the word refers to the transformation or deformation of bodies under the influence of some force. It evolves, etymologically, from the Greek *mechanastai,* which means something like "to devise a ruse," to produce some kind of artifice or substitution. The "ruses" in question were originally conceived as part of the machinery of theater: when Zeus comes down out of the sky on a cloud, the stage machinery and cables necessary to produce him participate in a deus ex machina, the god from the (theater-)machine. Leaving aside any interference from Galilean naturalization, we ought to understand the *mechanization of communication,* in this perfectly Aristotelian sense, as the attempt to assist or empower the spoken word through the use of a variety of technical mediators (devices, apparatuses, machines). Historical examples of the practice come easily to mind: Thomas Edison's phonograph (or gramophone) and Johann Philipp Reis's telephone.

It is today perhaps difficult to grasp the revolutionary character of these two inventions. We live in a technological civilization, surrounded by telephones and answering machines, by Muzak specifically designed to alter moods, by digital voice recorders or electronic diaries shaped like credit cards or ballpoint pens, and so on. Take, for example, a navigation system. Giving the driver acoustic instructions, responding to verbal cues, such a system constitutes a speaking and knowing machine, knowing the driver's location and the best possible route to a destination, becoming a basic commodity of everyday life. You have to imagine yourself back in the mid-nineteenth century, when the spoken word was still an ephemeral, and purely human, creation.

The phonograph, developed by Edison beginning 1877, made real a practice that had no correspondent in nature. To mechanically convert human speech into scratches on tinfoil, and to preserve it there, in order to reproduce this "recording" at a later time: this was a completely new, completely artificial idea. Through it ephemeral experience, experience whose very pastness seemed to define the transient and the fleeting, became technically available in any present, immediately ready to hand. The readiness, the availability, of speech no longer depended on the physical or temporal presence of the speaker. *The spoken word became a permanent thing, capable of moving through time and space.* The revolutionary character of this technological innovation is no less important than the much earlier apprehension of spoken words by writing. In fact, it supersedes it: what technology captures today is the authenticity of the speaker in a particular, irretrievable situation. Both forms of technical culture, writing and sound recording, completely revolutionized the experience of linguistic situations (I discuss writing at greater length in chapter 5).

It matters, then, that the apparatus that served to conserve and reproduce human speech was conceived primarily in mechanical terms. The electrical and electronic improvements leading to today's high-fidelity technologies have focused exclusively on the perfection of mechanical principles that treat sound-events as pure variations in air pressure, as spatiotemporal alterations in the physical world whose control and management can be produced, and used, in the recording of sound.

There is a striking similarity between Edison's invention and the telephone: just as the phonograph and its modern successors (in recording, magnetic cassette tapes and dictation devices, playback, record and CD players) eliminated the need for the speaker's presence and prosthetically replaced the immediacy of living speech, so the telephone (invented in 1836 by Johann Philipp Reis and patented in its modern form by the Scotsman Alexander Graham Bell some fifteen years later) came to stand in for the necessity of face-to-face communication. Though the two participants in a telephonic dialogue still needed to be in the same time, they no longer needed to be in the same space nor to share the forms of sociality, like eye contact, that belong to spatial copresence. What the telephone superseded was not the *speaker's speech* as such but rather the entire regime of ordinary, pretechnological interaction, the interaction between *communication* and *the personal presences and spatiotemporal concomitance of the people who were doing the communicating.* (To grasp this difference, it may be helpful to imagine why humans don't learn to speak by telephone: you would be missing the possibility of gesturing to events, people, and things, the entire armature of physical reference.)

It doesn't matter much here that telephone technology, like telegraphy, also developed in the mid-nineteenth century and did not get under way until the widespread advent of electricity. The precursors of mechanical telegraphy (semaphores, for example), or even a child's tin can telephone, indicate clearly enough that the transfer of speech across space is in principle mechanically possible.

The transition of words like *machine* and *mechanical* from their literal to their figurative meanings can be explained by their deployment in the fields of electronics and electrical engineering: *the mechanization of the spoken word and of communication* in the figurative sense means that communication can be causally managed according to physical laws and that its transportation across space and/or time can take place by exclusively technical means, and technical means alone.

It's to some extent psychologically understandable that, given the enormous technical power of the machines that would make the spoken word so available, no one would pay any special attention to the fact that the mechanization of speech could only take place as a result of its *reduction to noise,* to sound-event, and finally to variations in air pressure. Hence the story, reported shortly after Edison's introduction of the phonograph, that some early observers took the presentation of the apparatus for a magic trick, a carnival attraction, in which a ventriloquist spoke whatever words had been prerecorded into the machine. The same thing happened later on with the cinema and sound film, which were taken almost literally as moving pictures, that is, as photographs that had learned how to walk or speak. In each of these cases we witness the shocking apprehension of—or the failure to apprehend—technology's nearly unimaginable capture of a piece of specific, quotidian human life.

The specific instance of the revolutionary, unfamiliar dimensions of the mechanization of the spoken word leads to one of the major misunderstandings belonging to the more general *legacy of the mechanization of information,* namely, that the phonograph itself "speaks." (This is not merely a problem of the early twentieth century: consider again the way we ourselves treat the speech of contemporary car navigation systems and the degree to which the machine's speech and "knowledge" absorb something of everyday speech.)

At the limit, this mechanization of speech produces a false imaginary of the mechanical apparatus itself, one that treats it as *equivalent to human ability.* In everyday language we speak completely casually about the record player speaking, the calculator calculating, the camera seeing something, or the printer writing. But these ordinary words have become familiar, in human culture, foremost as expressions of human capacities; and these capacities in turn are taught, learned, and practiced in a variety of social and cultural forms. This means that the very words that culture takes up to describe the activity of its machines are the same ones used to describe the human capacities that these machines themselves replace (I mean "replace" in a specific sense; see "Equal Performance and Methodical Ordering" in chapter 5). This ambiguous, or simply double, usage stems entirely from the mistaken belief that human capacities like speaking, calculating, seeing, and thinking can be taken over and performed identically by machines.

Nineteenth-century developments in the physical, engineering, and technical sciences were paralleled by major leaps forward in the understanding of human sensory physiology, which investigated the visual and auditory systems as well as the various systems involved in the production of sound and speech. The success of this research, even today, lies in its capacity to develop functional models for the ways sound emerges from vocal cords, the roles played by the tongue, lips, and palate in the production of consonants, or the way the structures and functions of the inner ear participate in the activity of hearing; thus modeled, these processes can be explained causally. These developments reinforced the belief that sound and speech constituted a new form of the body–mind problem and imagined that such a problem could be resolved, or even mastered, by technical means. When the function of the human ear can be duplicated, and explained, by its reproduction in a complex microphone, then obviously you're going to end up understanding the spoken word itself in similar terms, following that same logical "channel" that got you to the microphone-as-equivalent in the first place. And likewise, the physical apparatus of speech must be the thing that produces the sense and meaning of spoken words . . . or so the misunderstanding goes, following as it does from the scientific successes of sensory physiology or the mechanization of speech.

At this point we can list a number of way stations along this path, like the hopes raised by the first enthusiasms for the new science of cybernetics (in the mid-1950s), which dreamed of explaining "speech" via physical descriptions of sound-events, or of discovering how words and sentences acquire meaning or come to be false or true. (This project launched a thousand schools of linguistics.) The optimism directed toward machine translation, itself part of a larger ambition to empty every human capacity of knowledge, thought, understanding, speaking, or acting into one or more imitation machines, belongs to this general trend. At the far end of these dreams lay the fantasy of artificial intelligence, of building hominid machines that would be indistinguishable from humans—and there of course we reach the moment at which the history of technology turns into (and takes over) the philosophy of human nature. The icon of this operationalization was Alan Turing's famous Turing test, which specified the conditions under which one could legitimately stop distinguishing between human and machine. At the moment when one can no longer decide, in the "information exchange" of "communication" between a test subject and an output-machine, which of the two is human, the naturalization of information passes back through the causal explanations of physics, and brings itself to term.

THREE
ARTICLES OF FAITH

New ideas, techniques, methods, and theories come, like new goals and their realization through scientific and technical progress, from people. These people have particular knowledge bases and knowledge gaps, tastes and aversions, and scientific and institutional biographies, particular self-images and world-pictures. All this is obvious and unavoidable. I have no intention of producing a psychological, much less psychoanalytic, interpretation of technoscientific innovation. I'm interested rather in the *explicit utterances* made by the authors of modern information concepts and information theory. Everything you need to understand the philosophical background of information appears in the plain ink of its public discourse, in all the clarity anyone needs. The inheritances that were the subject of the previous chapter function there, with immense historical effects, as articles of scientific faith.

The best-known and most influential information theory is Claude Shannon's (which appeared in book form in 1949). In an apparent chronological paradox, we need to discuss Shannon and his theory before getting to Norbert Wiener's 1948 *Cybernetics, or Control and Communication in the Animal and the Machine,* since Wiener's work refers explicitly to Shannon's.

Shannon's theory also cannot be understood without knowing that it was published alongside a philosophical explanation of it, written by Warren Weaver. This intellectual twofer was published by the University of Illinois Press in September 1949 under the English title *The Mathematical Theory of Communication*. The book's preface, co-authored by Shannon and Weaver, explains that Shannon's two-part article "A Mathematical Theory of Communication" had appeared in the *Bell System Technical Journal* of July and October 1948. The book version contained the article as well as a thirty-page discussion by Weaver (titled "Some Recent Contributions to the Mathematical Theory of Communication"). Weaver's essay had earlier been submitted to *Scientific American* and had appeared there, in abbreviated form, in July 1949. In this way Shannon's theory in book form appeared at the same time as the philosophical explanation of it. (I won't be discussing the various disagreements between Shannon and Weaver, because it was Weaver's philosophy that contributed to no small extent to the unfolding mythology of information naturalization.) Why and how the German translation of the book ended up making a telling change in the book's title—*A Mathematical Theory of Information* instead of *Communication*—is unknown.

It seems to have escaped general and scholarly notice that the philosophical background of these new approaches had been developed and published some fifteen to twenty years earlier, in the semiotic theory of Charles Morris. Morris had laid out, in an essay for the *International Encyclopedia of Unified Science,* an entire program for the unification of all scientific knowledge, grounded in the logical empiricism of the Vienna Circle. Morris's name did not appear, despite the numerous mentions of the work of a wide variety of others, in Shannon and Weaver's book. That makes the correspondences between Weaver's text and Morris's all the more astonishing.

The second important article of faith in the more general legacy of information theory, epitomized by the iconic claim that "information is information, not matter or energy," is Wiener's cybernetics, understood as a *mathematical theory of control and feedback technology.* The technological prehistory of that theory includes James Watt's invention of speed controllers for his famous steam engine, the development of electrical oscillators, and, finally, the rise of electrical network theories (knot and mesh theory applied to electrical circuits in the work of Hermann von Helmholtz, James Clerk Maxwell, or Gustav Robert Kirchhoff) and the development of electrical systems theory by Karl Küpfmüller in 1928. Cybernetics borrowed much of its intellectual language from pretheoretical control and feedback mechanisms, these latter developed some two millennia before the rise of electrical circuitry and having primarily to do with the building and operation of sailing ships. The intellectual and philosophical debt owed to those vessels and to their helmsmen (*kybernetes* in Greek) goes unacknowledged in contemporary histories of technology, which have no interest in their own legacy or development. Cybernetics drew its philosophical legitimacy instead from Wiener's Cartesianism and Darwinism, which organized living beings and machines under the same natural laws so as to be able to treat people as animals, and animals as automatons.

The third article of scientific faith, which appears outside the framework of any coherent intellectual–historical context, and ought probably to be regarded as an expression of historical irony, involves the mystery of *entropy,* a term that, because it was borrowed from the field of thermodynamics, stamped the idea of information (in Shannon's sense) with all the objective authority of physics. In this case, our understanding of the myth proceeds, somewhat surprisingly, not via an understanding of a major philosophical development and the errors it produces but rather by observing the historical force of a small ploy in the social struggle of fables and memes, a kind of sleight of hand of conceptual labeling, that probably has to be understood politically or psychologically.

These three articles of faith make up a curious mixture. On that mélange depends the contemporary sense of the fundamentally natural character of information.

Semiotics (Theory of Signs)

In 1938, the American philosopher Charles Morris (1901–79) published "Foundations of the Theory of Signs"; a year later, "Esthetics and the Theory of Signs" appeared in the journal *Erkenntnis.*[1] These two essays not only laid out the basic semiotic concept that would be used in Shannon and Weaver's information theory but also established the intellectual template for Weaver's essay, determining, at the limit, even its selection of examples.

It has not been much remarked how tightly Morris's texts and Weaver's correlate, nor how intensely their thoughts parallel one another's. Neither Shannon nor Weaver ever mentions Morris. This is all the more astonishing because the Shannon–Weaver text does refer to a number of other sources, and because it opens from practically its first page a contest between Shannon and Weaver on the question of who owes what to whom. The point here is not, however, to plumb the philological evidence to determine what Weaver owes to Morris but rather to remark on the more general relationship that ties the development of a naturalized concept of information to the logical empiricism of the Vienna Circle.

That's where Morris matters. As his introduction makes absolutely clear, his theory of semiotics was written under the aegis of the Vienna Circle's *International Encyclopedia of Unified Science,* where it aimed to supply, under the general framework of a unified theory, "a tool for the work of the *Encyclopedia,* i.e. to supply a language in which to talk about, and in so doing to improve, *the language of science*" (18, emphasis added). Morris then refers to his own essay on "Scientific Empiricism," published in the *Encyclopedia*'s first volume, arguing that he had shown there that "it is possible to include without remainder the study of science under the study of the language of science" (19).

CHARLES MORRIS'S APPROACH

Semiotics, the science of signs, takes its name from the Greek verb *semainein* (to indicate, to give a sign). Morris wanted to develop it, as a scientific field, from the field known as semiosis or semiology (with which it shared an etymology). He wrote,

> Men are the dominant sign-using animals. Animals other than man do, of course, respond to certain things as signs of something else, but such signs do not attain the complexity and elaboration which is found in human speech, writing, art, testing devices, medical diagnosis, and signaling instruments. Science and signs are inseparately interconnected, since science both presents men with more reliable signs and embodies its results in systems of signs. Human civilization is dependent on signs and systems of signs, and the human mind is inseparable from the functioning of signs. (17)

Here we see Morris taking up the typical Vienna Circle equation between knowledge and science *(Wissen and Wissenschaft)* and claiming to have produced, with a new theory of scientific language, the final inheritance of the entire philosophical tradition. At the same time he grounds his entire program in the naturalistic equation between humans and animals, reproducing the basic assumptions of biological science.

Morris begins the essay's main section on "Semiosis and Semiotic," under the heading "The Nature of a Sign," with an example from animal life:

> A dog responds by the type of behavior *(I)* involved in the hunting of chipmunks *(D)* to a certain sound *(S)*; a traveler prepares himself to deal appropriately *(I)* with the geographical region *(D)* in virtue of the letter *(S)* received from a friend. In such cases S is the sign vehicle (and a sign in virtue of its functioning), D the designatum, and I the interpretant of the interpreter. The most effective characterization of a sign is the following: S is a sign of D for I to the degree that I takes account of D in virtue of the presence of S. Thus in semiosis something takes account of something else mediately, i.e., by means of a third something. Semiosis is accordingly a mediated-taking-account-of. (19)

Notice that neither the dog who hears a squirrel nor the traveler who receives a letter contributes anything to the process; they *receive* signs. We are not in a situation in which actors actively produce signs but in the passive or receptive site of an event that requires a sender or initiator, that needs some kind of initial cause.

Unfortunately, Morris's text is (neither in the original nor in the German translation) not especially felicitous or exact in its selection of terms. For instance, there appears alongside the interpretant I "as a fourth factor" the "interpreter" (in Morris's examples, the dog or the traveler). As a result, it's useful, in the analysis of the philosophical premises of Morris's work, to return to a better-known illustration of this process, namely, the semiotic or semantic triangle developed in the work of Charles Sanders Peirce. Peirce differentiates in the process of signification among a sign (the mark or indicator of a meaning), an object (whatever the sign represents), and an interpretant (which mediates or translates the meaning of the sign). This last is not the *agent* (the traveler or the dog, in our example) but rather a fictive instantiation in which meaning is assigned to the sign. This instantiation can involve conventions of ordinary speech so that the idea of a house can be *"Haus"* in German or "house" in English but in Italian would need to be assigned to either *"casa"* or *"villa."* (Don't generalize on the basis of this example: these interpretants must also function for signs that do not refer to concrete things but to fictitious things, to relationships, or to objects of a completely different nature.)

THE SIGN-PROCESS BETWEEN NATURE AND ACTION

Morris's semiotics does not make it especially clear where the boundary of the sign-process lies when it comes to the position of the agent (or interpretant). As far as I can tell, it doesn't lie between the speaking person and the dog that instinctively follows its hunting instinct. It therefore doesn't seem to lie between linguistic or cultural performance, on one hand, and natural processes, on the other. But don't nonliving things also belong to the world of natural processes? Consider the following alpine situation: during a thunderstorm, a bolt of lightning hits a mountaintop, and the thunder that it produces triggers, on the slopes of another mountain, an avalanche. Would Morris describe the situation by saying that the tumbling snow has, through the intermediation of the thunder, "taken account" of the lightning? Is the "taking account of" a kind of *paying attention,* the kind of thing we use to describe what dogs or humans do? Is it a form of physical or biological *responsiveness,* the kind of thing that happens in plants, or is it some other, more general form of *causal transformation?* Obviously these questions are of real concern if we want to decide whether machines can take notice of things, or not.

Before I get to some answers, I want to pay attention to another of Morris's important contributions to the general mythology of information. Morris develops from the three objective positions (sign vehicle, designatum, and interpretant) three possibilities, and sets those possibilities and objective positions in relation to one another. This allows him to lay out three "dimensions and levels of semiosis": the *semantic* ("the relations of signs to the objects to which the signs are applicable"), the *pragmatic* ("the relation of signs to interpreters"), and the *syntactic* ("the formal relation of signs to one another") (21–22). But the syntactic relation of signs to one another cannot be determined within a singular, socially or logically isolated three-part sign-process (the semiotic triangle); it only can be understood (to speak geometrically) through the relation among many different semiotic triangles—which would allow you to understand what a piece of language within a system means in the first place. Though Morris writes that "this third dimension will be called the *syntactical dimension of semiosis* . . . and the study of this dimension will be named syntactics" (22), there is no need to follow him, no matter how appealing his conceptual generosity, into philosophical error.

We need to pay special attention here to questions of *ordering in the enumeration of concepts*. Some idea of the "natural"—some sense of the proper developmental order of the subject of investigation—underlies the choice to order semantics, pragmatics, and syntax in this way. We start with the primary relationship between signs and signifieds (semantics). As the examples of the dog and the traveler suggest, these relationships must be determined (in turn) by a ("pragmatic") *use-theory of signs*; they direct us to the reactions of the sign-receivers. One could sympathetically imagine, therefore, that every action of a given sign would bear alongside itself a kind of semiotic control group that would determine whether the interpreter (the dog, the traveler) had actually correctly "interpreted" the relevant signs. Syntax only emerges when a large context, an entire language or sign system, comes into view, bringing with it semiotic relations like those that, in the realm of human language, function more or less like definitions.

Morris abandons this "natural" order when, under the heading of "Language," he directs the reader to the one-sidedness that results when one considers one of the three aspects semantics, pragmatics, and syntax outside the operations of the other two. He writes, "The *formalist* is inclined to consider any axiomatic system as a language . . . ; the *empiricist* is inclined to stress the necessity of the relations of signs to objects which they denote and whose properties they truly state; the *pragmatist* is inclined to regard a language as a type of communicative activity, social in origin and nature, by which members of a social group are able to meet more satisfactorily their individual and common needs" (24–25, emphases added). Here the formalist, master of syntax, heads the list; the pragmatist takes up the rear. This new ordering of the analysis of "language," which serves to introduce an entire language of science, will be retained, and indeed canonized, from these pages forward, so that Morris will only speak of syntactics, semantics, and pragmatics, in that order, for the rest of the book. Today that ordering has acquired a status whose fixity can only really be compared to the force of the normative order of the numbers *one, two, three.* It determines the ordering of the steps in Shannon's information theory and has become a hallowed meta-norm for all analysis, whether in philosophy of science or philosophy of language.

It should not be overlooked that Morris directly follows the characterization of the formalist, empiricist, and pragmatist with these lines: "the advantage of the three-dimensional analysis [i.e., Morris's system] is that the validity of all these points of view can be recognized, since they refer to three aspects of one and the same phenomenon" (25). It's striking, however, that syntax comes first in the conceptual ordering only via a certain looseness and that the syntactic dimension cannot take place purely within a single semiotic act, or semiotic triangle, but only as an effect of the occurrence of more triangles, many triangles, indeed an entire system of signs. The elevation of syntax is thus the effect of a certain sleight of hand, whereby an apparently coeval part of a system becomes, instead, its origin and its ontological ground.

A CONVENIENT, METHODICAL ORDER

You don't have to look long to find the motivation and ground for syntax's elevation. In the first of three successive chapters, "Syntactics," "Semantics," and "Pragmatics," Morris writes, "Syntactics is in some respects *easier to develop* than its coordinate fields, since it is somewhat easier, especially in the case of written signs, to study the relations of signs to one another as determined by rule than it is to characterize the existential situations under which certain signs are employed or what goes on in the interpreter when a sign is functioning" (30, emphasis added). If we take Morris at his word, syntax takes its place at the origin of the philosophy of language because it's simpler and clearer to manage than pragmatics (which treats speakers in the context of their actions) or semantics (in which a speaker or listener stands between the signifier and the signified). Any understanding of how information came to be naturalized ought always to keep this explanation in view.

In the sentence that follows, Morris says something that will reappear almost identically in Weaver's work: "For this reason the isolation of certain distinctions by syntactical investigation gives a clue for seeking their analogues in semantical and pragmatical investigations" (30). (In Weaver the idea is first to resolve problems in telecommunication, problems involving the syntactic structures of large numbers of signals in information transmission, coding, and decoding, and then to present those solutions as motivations, or indeed as partial solutions, for the semantic and pragmatic problems of the field.)

The question I raised earlier, whether the sign-process ought to be understood as a kind of notice taking of human speech, of animal behavior, or of a simple physical, causal chain, finds an answer in Morris's next chapter, "Semantics." After deciding that neither linguistics nor the philosophy of grammar has gotten very far in clarifying the relationship between signs and what they designate, Morris continues, "*The experimental approach made possible by behavioristics* offers great promise in determining the actual conditions under which certain signs are employed" (35, emphasis added). Here the philosophy of language encounters, proleptically, the process that would be elaborated by W. V. O. Quine in his naturalization of epistemology.[2] Semiotics becomes tied to an experimental process and therefore to the possibility of *causal clarification*.

Another important moment in the mythical history of information appears here. That same sentence continues: "the development of the language of semantics has been furthered by recent discussions of the relation of formal linguistic structures to their 'interpretations,' by attempts (such as those of Carnap and Reichenbach) to formulate more sharply the doctrine of empiricism. . . . Nevertheless, semantics has not yet attained a clarity and systematization comparable to that obtained by certain portions of syntactics. Upon consideration, this situation is not surprising, for a rigorous development of semantics *presupposes a relatively highly developed syntactics*" (35–36, emphasis added). Here the *canonization and dogmatization of the order "syntax, semantics, pragmatics"* finds itself consummated and fulfilled. Having more or less casually installed the formalist ahead of the empiricist and the pragmatist, having taken the simplicity and researchability of syntax (by contrast to pragmatics and semantics) as the ground for their ordering, we arrive finally at the production of a strict prerequisitional relation between syntax and semantics, supported on one side by references to authorities like Rudolf Carnap and Hans Reichenbach (or Alfred Tarski) and, on the other, by the argument that any theory of semantics would have to be built on a foundation developed from a theory of syntax.

Thus are the mythological pillars of semiotics, on whose capitals "mathematical" information theory will construct its arches, laid down, and in such a way as to preclude semantic or pragmatic pillars from joining them. The consequences of this theory will come to fruition in the Shannon–Weaver theory. Their import lies in the investments and assumptions they silently bear and at which I wish, in conclusion, to take a final look.

When the word *communication* comes up in Morris's text, it does so in a context in which language is treated as a monologic expression of plain speech. (This reflects the Vienna Circle's general approach to language and would not be surprising in the logical tradition of Frege or Wittgenstein, either.) Because the Circle's main goal was to produce a logical analysis of scientific language, it had as its unconscious model of language practice the idea of a mathematician or a theoretical physicist sitting in front of a sheet of paper, wrestling with formulas, definitions, and proofs. Such a scenario treats language *(Sprache)* as little more than the elaborate end product of a purely theoretical practice *(Theoriesprache).* But in the cases of both individual language acquisition and the generation of speech, language can only be understood as a matter of social and communicative performance. This closes off all paths of entry to the semiotic aspects of language (in the order pragmatics, semantics, syntax). They must nonetheless *methodically* be walked: speech is a capacity for action that can only be meaningful, or be acquired, or practiced, in the context of other speech and other action and within the framework of human societies. To treat syntax as "simpler" than pragmatics or semantics involves a process of disassembly *(Degenerationsprozess),* a distancing, analytic orientation that allows its holders to take, case by case, the results of whatever recent intellectual developments in linguistics seem to them to be the most attractive as the starting points for their own thought. The dilemma this creates for logical empiricism and for Morris's semiotics results, finally, from a certain dogmatism. It sets into motion a mythohistorical inheritance from which information theory and its attendant concepts have still not recovered.

Cybernetics (Machine Theory)

Because in what follows I reckon cybernetics (with a side glance at robotics) as a mythology of the naturalization of information, I want to be clear about what the theory and practice of steering and regulation have to do with information or communication in the first place. This means taking into account the particular relationship between the kind of mathematics or computation that responds to the praxis and theory of operation or control (*steuern*, also "steering," as in the calculation of courses by navigators) and the ideas of information and communication.

CONTROL OR COMMUNICATION?

The obvious place to start is with Norbert Wiener, the father of cybernetics, whose groundbreaking book *Cybernetics* had the subtitle *Control and Communication in the Animal and the Machine* (1948; published in German in 1963 as *Kybernetik: Regelung und Nachrichtenübertragung im Lebewesen und in der Maschine*). Here we get not only the word *control*, pointing us toward steering and regulation, but also the word *communication*.

When one is on the trail of a prehistory of the naturalization of information, it helps to pay attention to key words. After all, the natural sciences also speak of communication in the case of simple causal interactions, as in the "communicating vessels" in classic hydrostatics. (In differently shaped vessels between which a liquid can be exchanged, the liquid will reach the same level in all vessels.) Because Wiener already in the subtitle of his book tells that he'll be speaking of both animals and machines, it would be a mistake to begin the reading with the hypothesis that "communication" here refers only, or even primarily, to language between humans—unless we understand that relation as more or less equivalent to the equalization of pressure in two water-filled vessels.

Wiener's book presents cybernetics, in a story that is largely autobiographical, as a mathematically grounded theory for technical control and management systems as it was first developed in the context of World War II military technology (at MIT). The military tasks begin with the control of antiaircraft guns, pass through the development of guided rockets, and end—as they do today—with the programming of cruise missiles.

Wiener aptly calls this approach "cybernetics," borrowing from the Greek word *kybernctcs,* "helmsman." A helmsman who moves a sailing ship from harbor A to harbor B guides it by the use of the wind and the ocean currents. The same thing happens in ordinary, essentially nontechnical contexts, as when a swimmer crossing a river accounts for the flow of its current if she wants to get to a specific place on the other side. One could therefore call "cybernetics" the *science of goal-following systems.* Should we wish to differentiate between systems that achieve steering in the narrowest sense and the organs of the execution of that control, we can use "robotics" (the word deriving etymologically from the Russian verb for "to work") as an additional concept: robots are (in the broadest sense) work-machines whose control systems the cyberneticist aims to develop. Because, therefore, robotics—regardless of any other definitions and classifications of that field—deals primarily with the technical implementation of control commands, we will not address it further here.

CONTROL AND REGULATION AS ACTIONS

I want to insist that it is philosophically relevant that processes that in engineering-speak go by "control" or "regulation" have as their basic conceptual and practical grounds activities of ordinary human life. What a rider does to her horse or the plowman to his ox both count as examples of the action of "control," because in every case the human determines the animal's path and objective. An even more ancient example of a control process is the simple maintenance of a fire, which can burn too hot or die out.

The so-called *control circuit* counts as a basic element of a general education within the framework of contemporary civilization. Nonscientists understand such circuits' basic principles and use them to control and regulate things. Common control circuits include the thermostatic valve in heating and cooling systems or the float valve that keeps a toilet tank from overflowing. Such valves save people the work of paying constant attention to maintaining the desired value of a temperature, the height of the water, and so on. What makes such mechanisms special has to do with their self-actualizing, "automatic" compliance with desired values, produced by the causal management of inflows and outflows of matter or energy. More naively, where something is *regulated* or *controlled* with the help of a control circuit, it must receive feedback from the actually existing conditions in a system to reconcile those conditions with the desired ones. This kind of "feedback" *(Rückmeldung)*, which etymologically recalls the idea of linguistic communication *(Meldung* is a message or a signal), must therefore begin with a sensor (a temperature gauge, a float) that *causally* influences an effector of some kind.

These trivial technical relationships become philosophically interesting if one notes that it takes a certain artificial effort, in any case, some kind of distance from ordinary conventions of language, to describe them without falling back on *cognitive* words like sensor or feedback (true or false?). This could be, however, as with the "communicating" vessels, simply a problem of the looseness of ordinary speech. Every regulation mechanism, from the flyweight governor of James Watt's steam engine to the quartz-stabilizing oscillating circuit of a modern wristwatch, can be *described in purely physical parameters,* without a single cognitive word, or indeed any word drawn from the quotidian realms of speech and action. Those physical parameters describe exclusively functional relationships.

THE DICTATORSHIP OF METAPHORS

As a first response to the question, What does cybernetics as a theory of control and regulation have to do with communication and information? we might therefore say the following: *this relationship involves merely a metaphorical, or partly metaphorical, description* of technical control and regulation systems. *On their own, the functions of control and regulation systems require no cognitive descriptions.* The special advantage of cybernetics over pre- or noncybernetic methods lies in its use of *feedback.*[3] But feedback for the regulation of a process itself is always *causally determined* and remains conceptually within the language and material framework of physics. To treat feedback as reflexivity (in the sense of the response of language to language, or thought to thought), by contrast, is already to metaphorize the concept. Metaphors like these do not add any new information about functional connections, which are already described by the language of physics.

The description of cybernetic systems thus operates via a certain metaphorical slipperiness. Terms like *target tracking, sensory control, intentional influence, feedback,* and *reflexivity* borrow their force from specifically human characteristics. This slipperiness originates, undramatically, in the simple fact that machines designed to regulate and control things replace *human capacities of action,* even if, in an important respect, they do so only partially: the selection of targets, the setting of goals, and the justification of values remain the province of the cybernetic system's inventor or user.

A second question—what role do calculation and mathematics play in the realm of cybernetics?—is not unimportant in the fields of informatics and computer science and is, by contrast to our first one, easy to answer. Consider, for instance, the calculations a navigator has to make regarding the speed of the airplane and of any crosswinds to fly from point A to point B. The movements for which the swimmer in the river or the shooter firing at clay pigeons make "allowances," leaving them to instinctual estimation, become linguistically more explicit in a calculation. That certain calculations, for instance, in highly complex electronic networks like the wiring of a television, require the entire scope of electrical engineering (and network theory) does not change anything about the basic relationships in principle: mathematics has no role other than the one it has in the calculation of the gear ratios in a transmission on the basis of the circumferences of the gear wheels. Mathematics is a tool applied to the construction of a system that is, in the final analysis, mechanical. A mechanical system, when it is cybernetic (that is, when it includes feedback), does not fall outside the scope of traditional physics. First, *certain machines are used for the achievement, maintenance, or avoidance of a goal, replacing human tasks in performatively equal ways.* These bring into play concepts that no longer belong to mathematics or physics. They focus on human interaction with machines and speak therefore of ends and means, of is and ought, of intentions, perceptions, and realizations. Only at that point does it become easy to understand why ordinary speech, somewhat sloppily, ascribes to machines the capacities of human beings.

CYBERNETICISM'S MATERIALISTIC MYTH

One might suppose, at first glance, that cybernetics, as a science of regulation and control engineering, could do with a minimal grounding in philosophy. Everyone is familiar with human efforts to influence or manage the course of events, whether in technical or pretechnical contexts of everyday life. The examples I've discussed so far—leading a dog, controlled maintenance of a fire, steering a sailing ship, and so on—should suffice to exemplify the setting and pursuit of concepts like ends, goals, and values. The engineer invents and discovers machines that allow people to let go of the burden of constantly monitoring and intervening in controlled and regulated systems.

Rightly seen, the transition to a natural science that investigates "natural systems" produces no philosophical problems. Natural processes will be "simulated," that is, functionally described and explained, by technical models. The homeostasis of blood sugar levels or body temperature is explained, for example, by analogy to a technical circuit like the thermostatic valve in a furnace. But the background philosophy familiar to formalistically oriented mathematicians, empirically oriented natural scientists, and natural science–oriented engineers leaves far behind a perspective that attends instead to the practice of daily life and to simple human activity.

I want to point, therefore, to two striking philosophical dogmas that appear in the work of Norbert Wiener, and in cybernetics more generally, and play there important roles in the naturalization of information. They are, first, scientific materialism and its consequences for the relations between human, animal, and machine and, second, an understanding of "higher," especially cognitive, achievements as "emergent" or "system predicates" of complex organized matter.

We've seen previously how Wiener's famous dictum "information is information and not matter or energy" and his claim that "no materialism which does not admit this can survive at the present day" (1948) concentrates the doctrines of this history of myths. Wiener's sentence articulates the information-concept's epistemic claim to be able to formulate scientific–materialist explanations, explanations that would otherwise, that is, as arguments about matter or energy alone, be impossible. (I leave aside any judgment of pseudo-biologism, of positions or programs that reduce all phenomena to material events by evaluating only their survivability rather than their truth.)

Wiener had meant that the *concept* of information ought to be placed alongside other scientific *concepts* like energy or matter. His formulation, which aims to compare information, energy, and matter as such, rather than differentiate the words or signs that indicate them (which would be to use "concepts" in a strict disciplinary sense), can be used only where its casual foreshortening will not lead to misunderstandings. The concept of energy belongs to the object language of the natural sciences, especially physics. Already in classical mechanics, energy operates in the same plane as "work" (force times displacement) and can appear in potential or kinetic form. In short, energy is a well-defined quantitative term in the theory of mechanics. All other forms of energy—thermodynamic, electromagnetic, chemical, and so on—are also measured by their mechanical or thermodynamic content and fall, therefore, within the framework of scientific quantities.

Things are different when it comes to the word *matter (Materie)*, which, in chemical contexts, is synonymous with *stuff (Stoff)* or *substance*. *Matter* does not indicate a quantitative concept and is therefore not defined by the operation of measurement. It does not function, in scientific discourse or science textbooks, as a term in scientific object language. It appears mostly in introductory or explanatory contexts that elaborate chemistry as a science of solids, their characteristics, and their changes. And where the word shows up in textbooks on its own, it has different uses, appearing both as "stuff" (*Stoff*, "matter," in the singular) and "elements" (*Stoffe,* in the plural). *Stuff (Stoff)* in the singular is, in truth, a reflection term, just like the words *space* and *time* in physics.[4] Reflection terms express metalanguage; they serve to classify and define facts formulated in object language.

Wiener's rhetorical carelessness involves treating his three terms as *words* instead of *concepts*. This makes it hard for him to see that he *asserts information's autonomy vis-à-vis very different objects of comparison*— vis-à-vis, that is, objects operating at two different levels, if one considers their definitions and actual use: *energy* belongs to the object language of science, *matter*, a reflection term, to its metalanguage. One therefore cannot tell whether Wiener wants to classify information as part of the former or the latter. On first glance one might say that the fineness of this distinction ought to be attributed no especial importance, so long as we respect Wiener's likely intention to give information an equal right of residence in the house of the natural sciences. As we'll see later on, however, this distinction as to the belonging of the information-concept to one or another level of language practice will serve as a reliable index of our major question here: whether information is, in fact, an *object of scientific practice* or *only a metaphor* that belongs to the realm of scientific modeling.

CYBERNETICS'S MECHANISTIC DOGMA

Wiener's materialism represents yet another major position in the mythology of information, inasmuch as it brings together ideas from Descartes and Darwin: Descartes's treatment of animals as automatons and Darwin's treatment of humans as a particular species, and hence as animals. By the miracle of casual transitive logic (human = animals, animals = automatons, therefore, humans = automatons), the cyberneticist gives us a new vision of the human and the world. It cannot be denied that it would be extremely useful to develop a technical science of goal-oriented machines that could provide models for the naturalistic aspects of animals and humans, provided that these in turn were described as organisms using machine models. Such a point of view only becomes problematic, a matter of dogma, when someone makes of it a universal claim and grants it the exclusive right to tell the truth about the world.

In the context of the natural sciences, "naturalization" has historically meant giving scientific descriptions and explanations the exclusive right to represent the real, a right that such descriptions and explanations would have over the humanities, the social sciences, or the study of culture. This brings another moment of the history of dogmatism into view: the reduction of humans to animals, and of humans and animals to organisms that can eventually be modeled by machine theory (certainly possible, and even useful in certain situations), passes, in typically materialist fashion, "from below to above," that is, from purely material events to higher ones, to the most complex manifestations of living things, of human beings, of intellectual and cultural life. The materialist conceptualization of the reduction of nonmaterial aspects to material ones appears in the idea of "emergence": the passage from simple to complex systems produces, as though from nothing, newer, higher system-characteristics, which we call—pre- or nonscientifically—"living" or "self-aware." (On emergence theory, see chapter 4.)

Here once again it's good not to be too impressed. Whoever would like to say—in a conventionally materialist and mechanistic way—that a mechanical clock, a system made up of parts, measures time and thereby fulfills a function that does not appear in any of its parts on their own must necessarily speak of an artificial object that can only appear in a world in which someone *first* had the idea of measuring time and *then* sought to discover and to create a technical means to accomplish that goal. In other words, no one needs to embrace the simpleminded idea that clocks are somehow there in world, like the moon (to repeat Galileo's misunderstanding of his technical apparatus), or to believe that, as the basic properties of the system's parts awaken its higher functions, the higher system-characteristics lose all their mystery: the parts of a mechanical clock are not "parts" in the sense of a division into parts. They are not the products of dismantling the clock. What happened is that someone came up with a plan to produce a mechanical method for measuring time and, on the basis of that plan, produced these apparent parts as *components,* that is, as pieces of the composition of the entire apparatus, which is then, and only then, artificially produced. Only at that point do the parts function as planned.

The mechanistic approach we see in Wiener (and in cybernetics more generally) thus speaks always of systems from the position of *description in hindsight.* This is all the more astounding because cybernetics as a technical science does not practice what it preaches. Like other cyberneticists, Wiener starts with a specific set of goals, namely, the conceptualization of the control of antiaircraft fire or a ground-to-air missile. Philosophically speaking, Wiener's openly materialistic and mechanistic approach allows him, *via hindsight (Nachträglichkeit),* to assume a false philosophy of his own field. The philosophy is false because it does not describe the actual procedures of the cyberneticists in the exercise of their own science. In short, the materialist–mechanistic philosophy of cybernetics is the product of the field's own misunderstanding of what it does.

On the subject of the role cybernetics played in the naturalization of information as a concept, it remains therefore only to say that it is aligned with the formalist, empiricist, and naturalist traditions of the philosophy of mathematics and natural science.

Things are otherwise when it comes to an aspect of the history of dogma that is simultaneously scientific and political, namely, the relation between information theory and thermodynamics, or between the concept of information and entropy. At their origin is not a philosophy but a coup de main, a daring usurpation, a (not necessarily unappealing) craftiness of sophisticated illusion making whose major ironic consequence was that even respected experts, in ignorance of the true history of the concept, would strive to provide a scientific foundation for what was essentially a mislabeled idea. We'll discuss the hows and whys of that history in the next section.

Politics, or How to Succeed in Science

To think beyond the clarity and truth of a theory, to consider its implementation or putting into practice, is a political act, an act of scientific politics. In fact, the history of science and technology has something organic about it: even before Thomas Kuhn developed his theory of paradigm shifts, it was known that the success and failure of scientific and philosophical ideas, concepts, theories, and processes depended neither on the regime of dispositions available to their creators nor on the realm of an objective world-spirit that would guarantee the stability of the truth. So there should be no objection to the idea that a scientific author like Claude Shannon would allow himself to be guided, in his selection of words, more by the possible success of his theory than by the adequacy and precision of their meanings. Which explains (partly) how it came to be that Shannon decided to name his famous measure of information "entropy." The decision, Shannon reported, followed some advice from the physicist John von Neumann: since no one knows what entropy is anyway, Shannon would always have an advantage in discussions of the concept.

The simple assignment of a name would normally be of little philosophical interest. But a formal similarity between two conceptual formulations (the measure of information, on one hand, and entropy in thermodynamics, on the other) allowed Shannon's baptismal act to unleash an extensive debate, one that ended with the protagonists on either side happily shouting, "Victory!" Meaning that they believed that the Shannon information-measure and thermodynamic entropy *were the same thing.*

These noteworthy events, whereby a scientific witticism became part of recognized scientific doctrine, must, one supposes, derive from some kind of principle. After all, no one doubts the scientific quality and usefulness either of the Shannon-measure or of the basic thermodynamic concept of entropy. In fact, there are structural similarities here. They do not belong, however, to *similarities between the realms of radio transmission and heat transmission* but to similarities in modes of contemplation, or better, to *similarities in the philosophical preconceptions* that motivate the comparison between the two areas in the first place. Accordingly, beginning with the historical fact of Shannon's christening, the next section takes up the general history of the information-concept's success, following it through the establishment of connections between information and energy and on to the deep causes of its doctrinal self-affirmation, which include a number of evasions, omissions, equivocations, and suspicious elaborations of equivalence.

STRAIGHT-SHOOTING SHANNON!

In 1971 *Scientific American* published an article called "Energy and Information," written by Myron Tribus and Edward C. McIrvine, on the topic of the relationship between Shannon's concept of information and thermodynamics. The article begins by noting that the formula Shannon used to describe the state of knowledge (in a receiver) before and after the receipt of a message was the same as what Rudolf Clausius used, in 1864, to describe the relation of the states of a system before and after the transfer of heat.

Then came the real splash: the authors reported that Tribus had, in 1961, asked Shannon about the reasons for his decision to name the measure of information "entropy." Shannon told him, "My greatest concern was what to call it. I thought of calling it 'information,' but the word was overly used, so I decided to call it 'uncertainty.' When I discussed it with John von Neumann, he had a better idea. Von Neumann told me, 'You should call it entropy, for two reasons. In the first place your uncertainty function has been used in statistical mechanics under that name, so it already has a name. In the second place, and more important, no one knows what entropy really is, so in a debate you will always have the advantage.'"[5] (In my translation of these sentences into German, I've used *Ungewißheit* to translate the English *uncertainty*, though one sees more commonly, as one does in the German version of Shannon and Weaver, a discussion of *Unsicherheit*. Part of the fuzziness of this whole debate lies in an *insufficient differentiation between objects of analysis and talk about objects of analysis*. In the usual linguistic unconscious of the natural sciences, something that is in truth a partial ignorance attributable to human beings who describe things gets assigned instead to the more or less likely conditions and patterns of the object of scientific attention, so that human ignorance becomes a property of things. Critically seen, however, in both information theory and thermodynamics, we are not dealing with states before and after heat or message transmission but with particular knowledge about those states. In German, therefore, I speak of *Gewiß* ["certainty," in the sense of something known, etymologically related to *wissen*, "to know"] and *Ungewiß* [uncertainty]. My other option was the word pair *sicher/unsicher* ["sure"/"unsure," but also "secure"/"insecure"], whose etymology vaguely suggests, and in the language of technical science too easily suggests—as in a cutout switch *[elektronische Sicherung]* or a security lock—that it describes a characteristic of something in the world rather than a characteristic of the description itself.)

I am not going to criticize people for considering the strategic advantages of naming choices in the natural sciences. A felicitous choice of words has played, as any number of historical examples will show, an important and even sometimes decisive role in the struggle for survival of concepts and theories. The two authors of the 1971 article rightly point out that Shannon's selection of the word *entropy* had caused intense interest among theoretical physicists. In passing, they mention Léon Brillouin, who, in a series of articles on the topic, had asserted the equality of the two entropy-concepts, joining entropy and negative information under the "negentropy principle of information."[6] A more recent philosophico-scientific introduction to information theory provides this summary: "Information entropy and thermodynamic entropy are formally identical. Both quantities are the same, when one considers entropy as potential information, and thus as a measure of possible microstates in a macrostate."[7]

I will not pursue here the many ways in which information theory and theoretical physics later became intertwined via the medium of quantum physics. It's enough to know that a name, chosen for strategic reasons, having been spoken by its maker, embarks on a career of its own. Shannon went so far as to designate, on von Neumann's advice, his information-measure with the letter H, so as explicitly to analogize the thermodynamic H-function developed by Ludwig Boltzmann. Was reason's cunning here at work? Is it possible that in this historical moment a brilliant researcher divined a truly sensational connection, that in the midst of the technical task of optimizing communication channels like undersea cables for telephone calls, a scientific constant would emerge, an information-measure for communication that would be independent of the communicative form (teletype, spoken language, Morse code), a constant that would be structurally identical, even identical in content, to a constant in the study of the flow of mechanical and heat energy?

Probably not.

A CLARIFICATION ABOUT STRUCTURES

Faced with the indisputable success of Shannonian information theory and thermodynamics, one should attempt, nonetheless, resistance to astonishment. Consider the following facts: in one field of application the topic is communication, messages, and information, though this range is clearly restricted by the technical problem of optimizing the transmission of "signals" across space and time. In the other the problem reaches beyond classical deterministic mechanics to (classical) statistical physics, which considers how mechanical and thermal energies interchange. In that arena, statistical analysis of the microscopic kinetic energy of individual molecules allows the macroscopic temperature of a volume to be observed and controlled. First appearances do not deceive: one theory concerns *the technical control of a human practice* (communicating and understanding); the other deals with *natural objects and processes* (so long as one is willing to ignore that "natural" here refers to objects brought into focus by a measuring, experimental, and machine-oriented physics and not to argue that they might be, instead, the artificial products of that technology itself).

Is this another case of metaphorization, whereby (as in Morris's own argument) no distinction is made between human speech, animal behavior, and mechanical stimulation and reaction? Or is this a metaphorization like the one in which the concept of communication extends scientifically downward to include the process of the movement of fluid through pipes or tubes?

Here things are actually a bit different. It's not a question of the opposition between the action of the communicators, or more precisely of intentional speech and intentional understanding, on one hand, and natural processes in the Earth's atmosphere or *natural-law processes* in a heat engine, on the other. In both cases the opposition is between *models*: before anyone can "discover" and make tangible structural similarities between information theory and thermodynamics, both phenomenal realms must first be transformed by a model-building process. In other words, in both cases, one must make the phenomena in question pass through an abstracting and philosophizing perspective; only then can one enjoy the structural resemblances they share.

In the case of Shannon's measure for information, this means accepting some strong premises. The well-known *Scientific American* article will stand in here for a host of other similar examples. There, for instance, a question Q is set up and opposed to a quantity of all possible answers, in order to designate (as the subject of a thought experiment) a question *for which the quantity of all possible answers is already known.* This suggests that, to understand questions, we will only ask questions for which the possible answers are already at hand. That's a pretty strange assumption, considering the role of actual questions and answers in human communication. From the point of view of the philosophy of language, questions are invitations to a partner in communication, essentially requests to remedy the imperfections of an incomplete sentence (either by answering yes, no, I don't know, and so on, for certain kinds of questions, or by providing the appropriate information for questions beginning "who," "what," "when," "where," "how," or "why"). This means that a *question from one person to another* can only be meaningful when the person who is asked the question can freely decide what to do in response, where the answer is open and not selected from a closed list of all possible answers, which the questioner has almost certainly not drawn up in the first place!

Where the quantity of all possible answers is unknown, you would have to interpret any given question as though it were a request for help in correctly formulating the question that would make the list of all possible answers available. Or at least that's what the authors of Shannonian information theory seem to believe.

Once all the answers are known, the problem is to figure out which answer (from all the possible ones) is correct. This produces two extreme cases: either one knows which answer is correct—and therefore also that all the others are incorrect—or all answers are equally possible, in which case the system "question plus the quantity of all possible answers" produces a maximum amount of ignorance or uncertainty *(Ungewißheit)*. Apart from such special cases, knowledge of the correct answer would always lie somewhere between 0 (it is impossible to know the answer) and 1 (some specific answer A is certainly correct).

This modeling mechanism, which in the account of Shannon's information-measure passes for "uncertainty" or "entropy," that is, the uncertainty in selecting from a *pregiven supply* of answers, or of signs more generally, is what builds the bridge that connects information theory to thermodynamics: with it we come, simply speaking, to models for the explanation of phenomena that are known to nonscientists. If one brings two bodies that differ in temperature into contact, the warmer one warms the colder one until the two arrive at the same temperature. This goes for solid, liquid, and gaseous bodies. If one follows a microscopic model—one that focuses on the mechanics of individual particles and molecules—one can imagine the two sides of a closed, isolated container (both sides having the same volume and pressure, but with their gaseous contents having different temperatures), in which the difference in temperature between the chambers will be managed by opening a partition between them. And now we arrive at the connection to information theory, because in this situation, thermodynamics too brings into play *the concept of knowledge,* describing the state of probabilities of the molecules in both halves of the container as "known" and "unknown." The knowledge of the empirically indisputable facts that, first, after a while, the number of molecules in both halves of the container will be the same and, second, that at no point will all the molecules gather in one side of the container develops precisely from what the model represents and not from some controllable system of measurement. It is *knowledge understood* through probability: one knows that the even division of particles between sides has a probability of 1 and that the segregation of all particles in a single partition has a probability of 0.

Both information theory and thermodynamic approaches tend to compare situations before and after a transition of some kind, the former before and after the sending of a message to a recipient, the latter before and after a change in temperature. Defenders of the thesis that Shannon's choice of "entropy" as a measure of information corresponds to thermodynamic entropy not only in form but *also in content* rely on the following argument from Carl Friedrich von Weizsäcker: "We have correlated information with knowing and entropy with ignorance, and as a result have referred to information as negentropy.

But this leads to a conceptual or verbal unclarity. Shannon's H is also the symbol for entropy. H is the expected value of the novelty-content of an event that has not yet taken place, and so a measure for what it is that I could know, but which I do not yet know. H is a measure of potential knowledge, and thus of a certain kind of non-knowledge. The same goes for thermodynamic entropy: it is a measure of the quantity of microstates in a macrostate. So it measures how much the person who knows the macrostate could still know, if he could also learn about the microstates. . . . What we call entropy . . . is the potential information contained in the macrostate. Entropy is greatest in the state of complete thermodynamic equilibrium . . . because in that state the amount of actual information in the microstates is at its smallest."[8]

For this comparison between two instances of knowledge and ignorance to be legitimate, several conditions need to be fulfilled: whereas the information model requires the presence of a predetermined, finite list of possible "answers" (or signs, that is, sign combinations taken from a finite stock of signs), in the thermodynamic example, the macrostate called "temperature" must be determined by a finite accounting of particles and their location in one or another side of the vessel. In both cases, entropy will be a relative measure of ignorance, the "expected value of the novelty-content of an event that has not yet taken place, and so a measure for what it is that I could know, but which I do not yet know." What's important, as von Weizsäcker correctly recognizes, is that this is a measure of a "certain kind" of ignorance, relative to a knowledge of the total quantity of possibilities corresponding to the thermodynamic macrostate.

The fundamental incommensurability of modeling information theory via a comparison to a model of thermodynamics does not seem to bother anyone much. It's not just that the common understanding of questions and answers is incompatible with the premise that one could have an overview of all possible answers for every single question. It's also the absurd way that the whole thing ignores the role questions and answers play in ordinary communication, inventing instead an *Archimedean observer* who watches and describes from an impossible outside the interplay between sender (questioner) and receiver (answerer) before and after the sending of a message, and who, from that position, attributes to the receiver an amount of previously unknown novelty-content theoretically transmitted by the message's means.

The astonishing success of this theory, implemented and enforced in part via a daring act of naming that produces an equivalence between two events—"a message goes from one person to another" and "an amount of energy passes from one quantity of gas to another"— carries with it serious theoretical premises that have determined the history of the dogma of information. They need to be considered in more detail. For the avatars of information theory, on what grounds does this successful equivalence finally rest?

THE DYNAMICS OF CARELESSNESS

The naturalization of information did not fall from the sky. It happened, rather, as the result of a single-minded, deliberate process whose goals make its ambitions clear: to reconstruct information as an objective, natural fact. (That it was a reconstruction cannot be denied; the entire process originates in human communication and its technological support by the telephone, the telegraph, and so on.)

The grounds of this reconstruction remain obscure. In its dark depths mingle any number of prejudices, supported by conceptual carelessness. I've addressed the first a number of times: that words, on whose meaning the success of communication between human beings depends, are translated (and used metaphorically) to discuss processes pertaining to plants, animals, or inanimate objects, all under the general heading of "communication." The second prejudice, closely linked to the first, is the refusal *to distinguish between human beings and nature* in terms of speech or action. The third, also tightly linked, is the refusal *to distinguish between the natural and the artificial (technical).* All artifacts—from the measuring instruments of physics misunderstood as a naturalistic discipline to the wide variety of machines that transmit messages and news to the feedback loops or computing machines of the cyberneticists—have their functions determined exclusively by their inventors, builders, and users, who aim to accomplish human goals. Gases in containers, plants exposed to sunlight, and even dogs that hunt squirrels: all these only *behave naturally,* that is, "causally in relation to natural laws," when they are studied by natural science.

The Paths of Error

Let's look back on the three scientific inheritances—(1) the naturalization of the sciences, (2) the formalization of theory, and (3) the mechanization of communication—and set them against the history of scientific dogmatism, that is, against the investments in (1) semiotics (a theory of signs as a philosophy of language), (2) cybernetics (focused on machine theory instead of the art of engineering), and (3) politics (after-the-fact rationalization of a feat of naming, instead of a separation of categories). Setting these two triptychs together opens onto a larger picture, a systematic context for the production of defects, lapses, and errors.

In individual cases the relationships emerge clearly. So, for instance, you find the Galilean misunderstanding of the natural character of scientific instruments repeated in the cyberneticists' machine-theoretical approach. The naturalization of the sciences that we saw in Hertz—the claim of isomorphism between nature and spirit—reappears in Morris's semiotics, which fail to distinguish between human and animal performance. The formalization of theory in David Hilbert leads directly to Morris's prioritization of syntax and affects in turn the restriction of Shannon and Weaver's mathematical theory of communication (or information) to so-called technical problems. The mechanization of communication via technical systems for the transmission and storage of human speech infuses all of cybernetics as well as the "political" equivalence of the transfer of heat energy and of messages. The equivalence of human and machine functions in the mechanization of communication shows up once again in cybernetics. And so on.

One can, instead of enumerating a series of contexts for the history of the reception of these ideas, do things in reverse, beginning with the specific defects of these processes and moving, as it were, upward: what they all share is a *misunderstanding of humans*. Humans are not seen as goal-directed, instrumentally rationally acting, and (as a particular form of that acting) speaking beings. Where, in Galileo, the experimenter's goal in the arts of experimenting and measuring was simply forgotten, and where, in Hertz, we find no distinction between the gains of historical and experimental experiences, humans become natural objects whose linguistic utterances degenerate into cultural trivialities. Hilbert's formalism, with its fixation on syntax, merely takes this process a step further, eliminating from the field of science any interest in the content of shared expression. The mechanization of communication conflates meaningful speech with natural sound-events; the world-orientation of meaningful language becomes, in Morris's behaviorism, a matter of instinctive responses to signals, governed by natural laws. In fact such responses are not really "responses" at all, just simple, undirected, natural events.

In short, any number of connections can be drawn among the history of the philosophy of science, mathematics, and technology, on one hand, and the history of the concept, theory, and technology of information, on the other. At the same time, it cannot be terribly surprising that the relationships among the elided differences, ungrounded assumptions, or doctrinal limitations of concern that I have complained about here do not allow themselves to be systematized in any strong sense. This is a peculiarity of errors, which cannot be seen from a coherent and systematic point of view. Mistakes are adjustments; even a doctrinal mistake is not a systematic and deliberate error.

Though it would be useful, then, for the sake of providing an overall view of things, to develop a logically and methodically ordered system of errors that would lay out the materialist, causalist, naturalist, realist, empiricist, and formalist positions, and describe the other modes of thought that fix and delimit philosophical opinions, such a project remains forever impossible. These positions belong not so much to a system of beliefs as to the general mood, a zeitgeist particular to the successes of empirical science and technology in the twentieth century. They serve to extend and strengthen a century-long debate about human self-understanding, one characterized by a belief in mathematics and physics that is so strong as *to forget the role humans played in their historical success.* The perspectival vanishing point toward which all this tends is a persistent *misunderstanding of the role* that acting and speaking human beings play in technical and scientific developments, first as *setters of goals,* second as *choosers of methods and means,* and third as *bearers of (legal and scientific) responsibility.* In each of the situations I've discussed, what's missing is the idea that one can, and must, confront the (philosophically misinterpreted) *achievements of mathematics and the natural and technical sciences* with the question, What for? These achievements have only appeared in the world because some human person brought them here. They are *cultural* achievements. The overarching framework of naturalism, which includes the syntactic–formal excesses of mathematics, logics, and semiotics, thus stands in opposition to a culturalist one that keeps the social life, historical background, and epistemological uncertainty of technical and scientific efforts fully in view.

So much for a provisional summary. It applies only to the intellectual inheritances, and to the history of dogmatism that they have influenced, which today ground the legend of the natural character of information. We now need to deal with the contemporary ruling ideas about information in the relevant sciences and technical disciplines. That is the task of the next chapter.

FOUR
INFORMATION CONCEPTS TODAY

The topics of information concepts, information theory, and the information sciences, as well as their respective technologies, media theories, and political applications, have produced an enormous outpouring of work, including their own academic journals, a scholarly society, and even, in 2006, a "year of information science." The sheer amount of material makes any overview difficult, especially if one wants to be intellectually responsible. There are some daring writers who have not shied away from the Herculean labor of providing such an overview, or who, knowing or feeling the difficulty of doing so, have nonetheless laid one out as part of a basic introduction to the field. Here I make no claim to any such completeness and provide no overview. I focus instead on a myth. Like all myths, it has a kernel, and this kernel explains how it came to be that information was mistaken for a natural fact.

Minimally, this myth needs to be a narratable history, so that its heroes and their actions can be identified. If the myth has more than one narrative form, taking place in everyday life, to which belong also the journalistic remediations of science and philosophy, or appearing perhaps in the mathematical, technical, or natural sciences, or again in the humanities and philosophy, its shape will vary according to the demands of its respective listeners and tellers. In any case, the myth of the naturalization of information appears everywhere. My own retelling of the contemporary understanding of information begins with its guiding principle, the *telecommunications paradigm,* for one simple reason: wherever a debate on these themes takes place, it returns consistently to a few elementary explanations that originate in the field of telecommunications. For this is where the clear examples of everyday life can be found, the ones we all know by virtue of our daily experience of relevant technologies like the telephone, the Internet, consumer electronics, and so on.

As a first step, we will look at what I call the "communications complex." It is (a) complex in two senses of the word: it is complex because of the entangled context that binds the standard example of the sender and the receiver to all that passes for reasons or causes, and it is *a* complex in the psychological sense, or to be more precise, in the psychoanalytic sense, according to which affectively intense experiences, though excised from consciousness, continue to manifest themselves in the form of mistakes or slips, producing communication without messages. Whereas the Greek loanword *technology* refers to an art and to something artificial, when it comes to communications *technology,* the human linguistic communication that it excludes is the means whereby the entire process crosses the ocean into the land of naturalism, on whose happy shores one hears only the chatter of neurons and molecules. This constitutes, I will argue, the most compelling form of communication today. Having said so, in a second step I will take this form as a basic explanation of the scientific understanding of information. At that point one will no longer be able tell whether it's the designers and copywriters or the scientific researchers who make everything—form and color, molecules and neurons—speak in the liquid voice of language.

Where the philosophical dogma has leveled all possible differences—those between humans and nature, nature and technology, action and function, or body and spirit—and where the grand unification of "structure theory for all" reigns, it will behoove us, in a third step, to discuss the ghost in the machine as it appears in the various relevant philosophies and sciences. Those three steps—the communications complex, the scientific understanding of information, and the ghost in the machine—structure the rest of this chapter.

The Communications Complex

Depending on your interests and points of view, you can hold different concepts of information: alongside the usual telecommunication theory (here represented by Shannon and Weaver's "mathematical theory of information/communication"), you could have an algorithmic one that would rely more on basic concepts in a number of different scientific fields: physics, biology, cognitive psychology, or some particular philosophy of mind. But all these concepts are naturalist ones. They all have as their basic objective reference examples that either literally are the classic example of message transmission or that can be represented in that fashion. That's the charm of the telecommunicative approach: it fits, especially in the naturalist mind, absolutely everything that you could ever mean by information.

MORRIS SAYS HELLO

The theory of *communication,* as it was called in Shannon and Weaver's original English title, or of *information,* as the German had it, pertains to the interaction between two objects that are called "sender" and "receiver." The theory's goal was to solve the telecommunications problem of quantifying the performance of a transmission channel that would (inevitably) be subject to disruption and to optimize its technical efficiency. No specific definition of "information" appears there (and indeed none was required). Rather, at least initially, Shannon chose for his measure of information a measure of effort (defined as a quantity of alternatives) drawn from the technically known means of message transmission (the classic example is the use of Morse code in radio). This involved selecting from a specific set of signs, like the letters of an alphabet, a concrete string of characters, like the words in a telegram. In later steps the theory transitioned from discrete to continuous messages, like the ones that appear in coherent spoken discourse of the type carried over a telephone connection.

The construction of the mathematical theory as Shannon put it forward at no point concerned itself with the idea that a given "set of signs" had to do with *signs,* that is, that these signs ought to be understood as part of a process in which a person "gives another person a sign" in an attempt to express or indicate something. But I am getting ahead of myself.

Before we allow Weaver, who wrote the book's first part, to take the stage, let's listen to what the actual inventor of the "mathematical theory of communication," namely, Shannon, has to say in his introduction:

> The fundamental problem of communication is that of repro-
> ducing at one point either exactly or approximately a message
> selected at another point. Frequently the messages have *meaning*;
> that is they refer to or are correlated according to some system
> with certain physical or conceptual entities. These semantic
> aspects of communication are irrelevant to the engineering
> problem. The significant aspect is that the actual message is
> one *selected from a set* of possible messages. The system must be
> designed to operate for each possible selection, not just the one
> which will actually be chosen since this is unknown at the time
> of design. (31)

Here, in a legitimate first step, Shannon sets his theory under a broader theory of conditions and prerequisites. Reading sympathetically, one might say that here "communication" is not *defined* in any comprehensive way but is developed under a specific program, namely, that of technical transmission of messages. The concept of a "selection of possible messages" would in such a scenario have nothing to do with communication that is human and free but would regard *only the functional limits of a technical system,* that of the transmitting machines.

Shannon goes on to explain why it's useful to use a logarithmic scale to produce a statistical measure for the arrival of individual messages:

> By a communication system we will mean a system of the type
> indicated schematically in Fig. 1.

> | Message | Signal | Signal | Message |
> |---------|--------|--------|---------|
> | Source → | Sender → | Channel → | Receiver → | Target |
> | | | Disturbance | | |

> The figure consists of essentially five parts:

> 1. An *information source* which produces a message or sequence of
>   messages to be communicated to the receiving terminal. The

message may be of various types: (a) A sequence of letters as in a telegraph or teletype system; (b) A single function of time $F(t)$ as in radio or telephony; (c) A function of time and other variables as in black and white television . . . (d) Two or more functions of time . . . —this is the case in "three-dimensional" sound transmission . . . (e) Several functions of several variables—in color television . . . (f) Various combinations also occur, for example in television with an associated audio channel.

2. A *transmitter* which operates on the message in some way to produce a signal suitable for transmission over the channel. In telephony this operation consists merely of changing sound pressure into a proportional electrical current. . . .

3. The *channel* is merely the medium used to transmit the signal from transmitter to receiver. It may be a pair of wires, a coaxial cable. . . . During transmission . . . the signal may be perturbed by noise. . . .

4. The *receiver* ordinarily performs the inverse operation of that done by the transmitter, reconstructing the message from the signal.

5. The *destination* is the person (or thing) for whom the message is intended. (33–34)

At this point Shannon connects the mathematical description of the individual components to their functions, dividing them into discrete, continuous, and mixed systems, first as undisturbed and then as disturbed processes.

The examples that Shannon places at the beginning and end of the transmission system, the information source and the information destination, are today such commonly known figures in the explanation of communications technology that they produce no particular astonishment—except of course for the fact that the *production of a message* or message sequence is described as the selection from a group of possible messages, as I've mentioned. What a message "is," or what it defines, remains unspoken. When it comes to the letters that make up a text, this idea is plausible at first glance and can be more or less translated to the situation of television, where we aren't dealing, for example, with a mass of possible images but with the limitations of the transmission-producing system, as when an image passing through the lens of a camera is registered as individual pixels on a scale of brightness values, according to the limits of the camera's resolution.

Now that we've established the general field of investigation that Shannon's theory takes on, it's time to hear from Warren Weaver:

> The word *communication* will be used here in a very broad sense to include all of the procedures by which one mind may affect another. This, of course, involves not only written and oral speech, but also music, the pictorial arts, the theatre, the ballet, and in fact all human behavior. In some connections it may be desirable to use a still broader definition of communication, namely, one which would include the procedures by means of which one mechanism (say automatic equipment to track an airplane and to compute its probably future positions) affects another mechanism (say a guided missile chasing the airplane). (3)

Here specifically human forms of communication appear, somewhat disconnectedly, next to mechanistic ones—what can it mean to say, in the case of a missile following an aircraft, that "one mind may affect another"? The lack of any transition from human to machine promises some excitement in Weaver's upcoming analysis, where he separates three levels of the communication problem by asking three questions:

> LEVEL A. How accurately can the symbols of communication be transmitted? (The technical problem.)
> LEVEL B. How precisely do the transmitted symbols convey the desired meaning? (The semantic problem.)
> LEVEL C. How effectively does the received meaning affect conduct in the desired way? (The effectiveness problem.) (4)

He follows this with explanations of the three problem types: under the technical problem, he puts the quality of the transmission of character strings from sender to receiver, as it is affected by disturbances in the communication pathway. The semantic problems, he writes, "are concerned with the identity, or satisfactorily close approximation, in *the interpretation of meaning by the receiver,* as compared with *the intended meaning of the sender.* This is a very deep and involved situation, *even when one deals only with the relatively simpler problems of communicating through speech*" (4, emphases added). Then we get a trivial example of a conversation between Mr. X and Mr. Y, who don't understand each other properly. Weaver's explanation of the situation is profoundly naive. Explanations (in a dialogue designed to clarify the first conversation) are "presumably never more than approximations to the ideas being explained," but they are "understandable since they are phrased in language which has previously been made reasonably clear by operational means. For example, it does not take long to make the symbol for 'yes' in any language operationally understandable" (4; the reader may want to look back at the discussion of the naturalization of physics in Heinrich Hertz, for whom language also has to do with thought-pictures of natural objects rather than with the relationships between objects and their description in language).

One is inclined to overlook the subsequent example, in which "the meaning of an American newsreel for a Russian" (in 1949!) illustrates the difficulty of semantic problems, in order to get to the problem of effectiveness: "The *effectiveness problems* are concerned with the success with which the meaning conveyed to the receiver leads to the desired conduct on his part" (5). These remarks of Weaver's are both conceptually lazy and confused with respect to the underlying model: weren't the *sender* and *receiver* distinguished from the message origin and message destination? Weren't they only the machines that translated spoken messages (as by telephone) into electrical currents? Or do machines now have intentions? Wasn't the receiver supposed to receive signals, not messages, which it was then supposed to turn into messages? Honestly, you really could almost not do worse, in terms of conceptual sloppiness, than what Weaver has done here. He ignores his very own terminology, which he's taken over verbatim from Shannon's system of five components and their functions. All this leads one to suspect that what really interests Weaver is only the three "inner" components of the system, that is, the sender, the channel, and the receiver; nothing here really concerns the production and interpretation of messages. *Only the coding and decoding function under noisy conditions count as the matter of the message, and thus as the subjects for a theory of signal transmission. Even the fact that these are "signals" (from the Latin* signum, *that is, "sign") ceases to matter under these conditions.*

Weaver follows this up with a (surprising) step:

It may seem at first glance undesirably narrow to imply that the purpose of all communication is to influence the conduct of the receiver. But with any reasonably broad definition of conduct, *it is clear that communication either affects conduct or is without any discernible and probable effect at all.*

The problem of effectiveness involves aesthetic considerations in the case of the fine arts. In the case of speech, written or oral, it involves considerations which range all the way from the mere mechanics of style, through all the psychological and emotional aspects of propaganda theory, to those value judgments which are necessary to give useful meaning to the words "success" and "desired." . . . The effectiveness problem is closely interrelated with the semantic problem, and overlaps it in a rather vague way; and there is in fact overlap between all of the suggested categories of problems. (5–6)

A sympathetic reader wishing to adjust these lines so as to make them compatible with the basic communication model that grounds them would *redescribe the information source as an intention and the destination as a reaction.* That beautiful art and aesthetic judgment, emotions, and worth come into play here is a sign that communication for Weaver comprehends not only trivial quotidian dialogues but the entire cultural field of human interactions. But how is the ambition of this high standard to be understood in terms of level A that Weaver described earlier, namely, in terms of the mathematical structures of signals in telegraphy, telephony, and television?

Weaver explains in "comments" that appear in the following lines: "So stated, one would include to think that Level A is a relatively superficial one, involving only the engineering details of good design of a communication system; while B and C seem to contain most if not all of the philosophical content of the general problem of communication" (6). After confessing that Shannon's theory, which follows, relates only to level A, Weaver nonetheless argues that "the theory has, I think, a deep significance which proves that the preceding paragraph [about the superficiality of level A] is seriously inaccurate. Part of the significance of the new theory comes from the fact that levels B and C, above, can make use only of those signal accuracies which turn out to be possible when analyzed at level A. Thus any limitations discovered in the theory at level A necessarily apply to levels B and C. But a *larger part of the significance* comes from the fact that the analysis at level A discloses that *this level overlaps the other levels* more than one could possible [*sic*] naïvely suspect. Thus the theory of level A is, at least to a significant degree, *also a theory of levels B and C*" (6).

These lines make an important promise: after the somewhat trivial remark that the poor quality of a technical intermediary might restrict the understandability of a message (who would have thought?), there follows a completely untrivial proposition that the technical problems of efficient signal transmission can resolve questions from semantics and pragmatics, questions about, in other words, the recognition and understanding of speech by a listener in the field of human language. The reader will search in vain for any attempt to ground this proposition in the philosophy of language. It would seem to be implicitly justified by the carrying out of Shannon's communications theory program. Or maybe the underlying philosophy of language functions here as an uncontroversial tacit premise.

So much for the telecommunications paradigm in Weaver's self-evaluation. But to what extent can we recognize in that evaluation the presence of those dogmatic inheritances that constitute, as we saw, a kind of prehistory of the naturalization of information? Here we need to trace the striking resemblances between the mathematical theory of communication and Morris's semiotics (some of which are implicit already in their titles, as we shall see).

In Weaver, communication seems initially to be limited to carriers who think (and thus can be part of "one mind" affecting "another") and have the capacity to be imaginative, that is, humans ("all human behavior"). How then does art show up to the party ("music, the pictorial arts, the theatre, the ballet"), how does "the problem of effectiveness" lead to "aesthetic consideration in the case of the fine arts" (5)? What motivates the leap from spoken or written language to "all the psychological and emotional aspects of propaganda theory, to those value judgments which are necessary to give useful meaning to the words 'success' and 'desired'" (5)?

As it turns out, these references to art, emotions, and values function, for both the theory and the book more generally, as *foreign bodies*. Language as the most important example of communication is described using statistical means and the Markov process and treated as a process of selection "with a view to the totality of things that man might wish to say" (27)—hardly an appropriate view for the discussion of art, emotions, and values.[1] We quickly find a solution to this oddity in Morris's "Esthetics of the Theory of Signs": for Morris, aesthetics is the science of artworks, insofar as these are understood as aesthetic signs. "Esthetic analysis then becomes a special case of sign analysis. . . . Esthetics in turn become the science of esthetic signs."[2] But semiotics doesn't cover everything, he writes: "if the theory of signs is one base upon which to erect the proposed type of esthetics, the theory of value is an equally necessary base—since it will be held that the designate of esthetic signs are values, or better, value properties" (418). (Morris's theory of value borrows from the work of John Dewey and George H. Mead and is explained using the example of the relation between the nutritional value of a food and hunger.)

This subordination of aesthetics to semiotics plays no significant role in Shannon and Weaver's information theory whatsoever; indeed, it contradicts their theory's purely syntactical scope. For this reason, Weaver's strange expansion of the concept of communication to include art, emotions, and values may be little more than a clumsy gesture toward his ambitious, totalizing reach (a reach to which Morris's claim to have mastered the entire of field of philosophy also testifies).

Recognizing that these theoretical foreign bodies in Weaver point us to a hidden reference to Morris allows us to sensitize ourselves for other aspects of the Shannon–Weaver theory, aspects that are in no way "external" to it. When Weaver lays out the problem of a "communication system" (made up of the five components of information source, transmitter, channel, receiver, and destination), he initially puts the components into the well-known examples of telephony and telegraphy. He then says, "In oral speech, the information source is the brain, the transmitter is the voice mechanism producing the varying sound pressure (the signal) which is transmitted through the air (the channel)" (7).

Here the shift from linguistic communication (in which the human speaker, a person, is the information source) to an organismic theory (that requires scientific causal explanations) is quite clear. Weaver's treatment of his own distinctions is, to put it mildly, somewhat generous: the vocal cords encode the neural message from the brain into spoken speech. How are nerve impulses a "message" and not signals?

On the receiving side, "when I talk to you . . . your ear and the associated eighth nerve is the receiver" (7). The same arbitrariness that allowed Morris to leap from nature (the hunting dog) to culture (the receiver of a letter) rules Weaver here. Everything comes down to causally explainable "behaviors," in Shannon, Weaver, and Morris alike. Semantics and pragmatics, and hence the meaning and value of spoken or written language, are no longer limited to anything specifically human. Even the "intentional behavior of a sender" belongs to the realm of natural facts. Intentions belong not only to humans and animals but also to machines. The key concept of Morris's semiotics, the *mediated taking-account-of*, becomes a *mediatedly influenced-through*. Whether something is a sign for something depends only on its attribution by an observer external to the entire five-component system. And whether that attribution is correct depends, in turn, only on the adequacy of the causal relationships inside the system. In other words, in Shannon and Weaver, one finds, in every detail, the same equation of objects that can bear meaning and value (as linguistic objects do) with natural facts and processes that we saw in Morris.

Above all, what emerges is the claim, unachieved by either approach, that the resolution of the syntactical or technical issues leads directly to the resolution of the semantic and pragmatic issues that follow from them (even if neither Morris nor Shannon–Weaver fully follows through on this project). Where, for Morris, the "rigorous development of semantics *presupposes a relatively highly developed syntactics*" (36), in Shannon and Weaver, limitations on the technical level affect semantic and pragmatic concerns. Where, for Morris (following the theory of logical empiricism), theoretical problems of meaning (semantics) and value can be solved (partially) through syntactic means, in Weaver it becomes clear that the analysis of technical problems "overlaps the other [semantic and pragmatic] levels more than one could possible [*sic*] naïvely suspect" (6). As in logical empiricism through the late works of Carl Gustav Hempel, and even in the seemingly oppositional position taken up in the philosophy of Karl Popper, semantic and value problems in the mathematical theories of physics find, at the end of the day, syntactic resolutions. Likewise, in Weaver, the theory that solves technical problems "is, at least to a significant degree, also a theory of levels B and C," namely, the semantic and pragmatic sides of communication (6).

(I am not going to pursue here the question of how direct or indirect the connection between Morris and Shannon–Weaver actually was. That would be a task for a historian of science. What matters here are the stunning parallels between the two underlying philosophies in question, whose trace appears in the spurious references, in Weaver, to art and values as "signs.")

"CAN YOU HEAR ME NOW?"
TELECOMMUNICATIONS' SEMANTIC DEUS EX MACHINA

To attend to and solve the technical problems of the transmission of messages does not require one to abstain from developing a philosophy of language that would deal with the classical questions regarding the meaning and value of linguistic utterances. From a philosophical perspective, there is nothing to criticize about the endeavor (it has been extraordinarily technically successful). But things are otherwise when it comes to the reduction of human language to telecommunicational structures, that is, to the temporal patterns and processes of physical effects and the taking of these latter *as complete solutions to questions of meaning and value.*

Unfortunately, that's exactly what Shannon and Weaver's "Mathematical Theory of Communication/Information," and indeed the whole tradition that follows the paradigm it establishes, does. Where, as with Morris, syntax takes disciplinary precedence over semantics and pragmatics, or, with Weaver, the problems of physical transmission between a coding and decoding machine become a theory of meaning and value of communicative acts "to a significant degree," the scope of language philosophy remains suspiciously unfulfilled.

A certain ideology has thus opened a path. That path originates, undeniably, in the use of the means of formal logic for the analysis of the language of mathematics and physics. That analysis, in turn, enacts a theory (which is itself the product of a great deal of intellectual effort). And this latter imagines an *enclosed system of language,* to which the syntactic and, following it, the semantic and pragmatic analyses finally refer. This prehistory grounds the later understanding of speech in the telecommunications context, treating it as the *mere selection* from a pregiven quantity of signs or sign combinations and describing it statistically as a Markov process.

It doesn't take a language philosopher to see two things in an initial glance. First, actual human speech does not involve the selection of words or sentences or text from a pregiven repertoire; that kind of selection process happens, maybe, when you use a telephone book or a rhyming dictionary. Second, the reduction of language to syntax or signal-structures makes it impossible to ask two (seemingly relevant) questions: (1) how do two communicating beings mutually understand each other? and (2) how would you ever test that understanding?

It is now possible to gather together the mistakes of this entire approach. They involve, first, its basic understanding of language as *monologistic*; second, the assumption of the existence of a naive external perspective, lying outside the processes of communication, when it comes to questions of semantics and pragmatics, that is, for questions of the meaning and value of verbal utterance; and third, talking about signs and signals (not to mention the various psychological contexts and circumstances that surround language, including the influence of mental images or the desire to produce an effect on a receiver), while describing what are in truth exclusively scientific or technical processes that might best be defined as successions of causal effects.

Let me explain. (1) The understanding of language is *monological* because the smallest unit of the transmission of a message from origin to destination, the equivalent of a telecommunicative "atom," does not include any anticipation or effect of the receiver's expected (linguistic, sender-directed) *response*. As the examples of the telephone and telegraph make clear, we're not talking here about real telephonic communication in its commonly understood sense (which would include speech and counterspeech, that is, *dialogue*) but rather about technical equipment of the types used to make announcements on train station platforms. Someone says something on one side, and on the other side a voice comes out of a loudspeaker. Travelers on the platform have no apparatus with which to respond to the announcement. As for the telegraph, it consists (in the examples) of an operator who uses a keypad to enter the signals that translate a telegram into Morse code. On the other side, there is only a receiving machine, like a pen that marks dots and lines on a band of paper. No one imagines an apparatus that would send back some kind of answer.

Naturally, the theorist of communication would gladly recognize that this is so and willingly expand the system to include the components of sender, channel, and receiver, now applied to the people and places in the opposite direction. This answer is good enough, on technical grounds. But as a description of the process and concept of communication, it simply will not do. That's because it requires us to believe—and this is why I say that its understanding of language is "monological"—that the problem of understanding between origin and destination either *does not or cannot take place* or that it cannot be resolved by the communicators themselves but only by an external observer. This lucky fellow would describe the impact of misunderstanding on the message destination with all the distance appropriate to a (behaviorist) natural scientist, whose semantic and pragmatic competence would allow him to explain the relationship between that misunderstanding and whatever happened at the message origin.

On to (2): this monological model of language is a consequence of the resort to behaviorism in Morris or of the reference to semiotics in Weaver. Together these produce the classical caricature of the two behaviorists. They do not greet each other by asking "how are you?" because this would require of each individual an impossible degree of introspection, semantic competence, and pragmatic orientation; instead, they ask "how am I?" because only the other of the speaker can, from a "scientific" outside, observe and resolve the question of how the speaker is doing, what he would like to say, what his utterances mean, and whether those utterances are successful.

Obviously the pre- and extratechnical principles that govern communication work otherwise, as do those governing communication via technical means like telephones or telegraphs. It's always the communicators themselves who judge and resolve the questions of meaning and value of what is communicated. No deus ex machina descending from the clouds, no Archimedean observer or scientific behaviorist from beyond the world, no God's-eye view, comes to confer meaning and value upon technically transmitted, physically determined transmission processes and effects. In every circumstance and every context, even when it comes to the evaluation of the philosophies of language and theories of information and communication, it is the case that language-using humans themselves, as speakers and hearers, produce and understand the meaning, value, and coherence of their linguistic acts and use pragmatic and semantic components to do so. That's how they regularly exchange their roles as speaker and hearer.

As for (3): in telecommunications it is strictly illegitimate to speak of "signals" or "signs" or "character sets," because we are dealing here only with the structure of physical parameters as they appear historically in space and time. Electromagnetic waves, for example, have structures, which can be produced by a loose contact in a defective plug or an erratically running motor in a refrigerator. These can be taken for signals, for instance, in the search for a problem by a repairman, who would perceive the irregular rattle of the fridge, hear crackling in a socket, or smell something burning. But only a purpose-oriented, understanding receiver of spatiotemporal changes in physical parameters can turn a clatter or a stink into a signal. The same is true on the sender's side of the telecommunicative paradigm. The goal-directed and purposeful invention, construction, and development of machines (in the context of intentional address, speech, and communication between humans) is the *necessary condition* that allows us to talk about signs, character sets, and signals. If, on the other hand, rain falls through a leaky roof onto the keyboard of a working computer, triggering some kind of "command" that, if the PC is connected to the Internet, has some actual effect in the world, we're still not dealing with signals or signs. Anyone who described things otherwise would presumably have to be willing to claim, absurdly, that a mouse "calculates" when it runs over the keypad of a calculator, thereby triggering some change in the signs on the machine's display.

These examples go so far as to prove that the essentially formalist and materialist approach of the logical empiricist tradition, like the materialism of the cyberneticists, requires a deus ex machina, a conceptual outsider, to get from the technical apparatuses and their causal effects, on one hand, to linguistic meaning, semantic recognition, or even influences on "mental images," on the other. In chapter 5, a methodical and systematic approach that repairs the errors we have seen so far will show that meaning and value do not just randomly emerge from physical patterns ("more than one could . . . naïvely suspect") but that, even so, human dialogue that produces meaning and value in relation to an addressee can in fact be reduced to its physical aspects, in the service of expanding human communication through technology by making it transportable across space and time.

ALL DONE? TELECOMMUNICATIONS' PRAGMATIC DEFICIT

And so telecommunications' deus ex machina is a semantic sham: using words like "sign" or "signal" makes it seem as though coding, transmission, and decoding machines have something to do with the meaning and value of human communication. But this is not the only mistake, and not the only sleight of hand that grants the theory a philosophical and social scope that it does not actually have. In both the language philosophy privileged by Morris and the telecommunications paradigm that organizes Shannon and Weaver, the same issues that make a muddle of semantic issues appear once again when it comes to the pragmatic realm.

I showed earlier how, in Morris's representation of semiotics, syntactics emerged as the leader of the trio syntactics, semantics, pragmatics (first via a casual mention, then in a simple argument, and finally as a conceptual prerequisite for the entire system). The communications complex of the theoretical tradition inaugurated by Shannon and Weaver continues, even in our contemporary communication, information, and entertainment technologies, to deal only with "syntactic" machines. However fancily or intelligently programmed, using "fuzzy logic" or creating "neural networks," whether digital or analog, these machines feature only causal reciprocity between switches, whether these latter follow the principles of a simple on–off mechanism (digital) or a lamp dimmer (analog).

It would be difficult to exaggerate the degree to which the entire debate about information concepts, theories, and technologies relies on retroactive, analytic, top-down argumentation. The cry of the army of technicians, scientists, philosophers, and journalists that have taken on this conceptual battlefield would seem to be "All done!" What's done, what's laid out in front of us as given or assumed, includes the logical, mathematical, and scientific–technical theories that find their way into information and communication theories and technologies; human language, with its syntactic, semantic, and pragmatic structures; language philosophy and its canonical–dogmatic orientation toward syntax, which places it before semantics and pragmatics; the idea that systems theory comes before systems; that systems come before components; that the reflex mechanism of a squirrel-hunting dog comes before its description in a semiotic theory; that the machines called "sender" and "receiver" come before the desire of one human being to say something to another. In the backward world of the communications complex, all conceptual ordering is *reversed*.

Everyone in this world knows, attends to, and practices this reverse ordering, even when it contradicts her own sense of things. Naturally, a person *first* decides to say something to someone, *before* she can arrive at the question of whether the circumstances require sending her a letter, recording a message on tape, or using the telephone or telegraph. Naturally, a person *first* learns how to act in the world (including how to use nonlinguistic actions appropriate to his culture and its various codified action sequences, such as the basic tasks of daily life, including eating, getting dressed, living, taking care of others, and so on) *before* inventing, building, or using telecommunication machines. Naturally, a person *first* learns to speak (in communicative situations) *before* delivering a self-directed monologue that works out a mathematical proof. Naturally, there *first* have to be functioning, that is, accomplished and successful, communications, in which the participants regularly switch roles between speaker and listener, and naturally, such a scene *first* requires an understanding of the meaning of what is said ("functioning communication") and a recognition of its shared meaning ("successful communication"), *before* the structure and noise of a single transmission from origin to destination (an utterance directed by one person to another) can be analyzed. Naturally, there *first* has to be a plan for a complex artificial construction (like a telephone system) *before* its technical realization can take place, and naturally, the technical system has *first* to be built *before* it can start to function; naturally, the system (like a telephone) must *first* function *before* it can be used. And naturally, the possibility of this entire sequence of steps can *only in retrospect* be described from a top-down perspective.

Though the communications complex's pragmatic deficit begins, then, with a methodically ungrounded ordering of syntax, semantics, and pragmatics, it is the product of a more general misunderstanding of the fact that the entire world of objects that we are discussing here would not be *available* were it not for human action. Decisions about functioning and nonfunctioning *(Gelingen und Mißlingen),* success and failure *(Erfolg und Mißerfolg),* happen in the realm of human action.[3] Only from there can one move on to technologies or theories.

You can sum up this analysis of the semantic and pragmatic deficits of the communications complex in one (admittedly long) sentence: the dogmatic division of theoretical philosophy in the logical empiricist tradition (via the amputation of pragmatics, particularly of the latter's productive, constructive elements) and the emphasis on analytic, that is, resolution-oriented methods, led its followers to believe that only the ready-to-hand products of science and technology, and not the human endeavors and reasonings that led to them, would permit an accurate understanding the world. Which is, after all, a pretty fundamental mistake.

Philosophical errors are, history teaches us, often as dangerous as the errors of scientists or technicians. (More gently, one might say that the dangerous errors made by scientists and technicians are in fact of a philosophical nature and are dangerous as such.) They have far-reaching consequences, affecting virtually all aspects of human perspectives and world-pictures, penetrating into every corner of private and public life, including politics and, indeed, the entire history of culture. We'll discuss those consequences in the next section.

The Chatter of Molecules and Neurons

It's easier than you might think to find molecules that speak, or conversations among neurons; prominent examples appear in any number of public discussions of molecular biology. From the moment of the announcement, made by President Bill Clinton and the entrepreneur J. Craig Venter at the White House in 2000, of the "unlocking" of the human genome, the media fell in love with a series of specific metaphors. The use of linguistic and cognitive metaphors to describe molecular and neuronal interplay has since then become so common, its penetration into our ordinary daily language so deep, that it hardly needs exemplification; new examples appear in the papers daily.

This trend is not exclusively an effect of science journalism. The intensity of this imagery does not stem from the well-known mediatic task of sharing with epistemologically impoverished laypeople (on one side of the journalist) the elaborate complexity of scientific concepts (on her other side). The overfreighting of cognitive and linguistic metaphors appears both in *everyday language* and in the language of the *scientific disciplines themselves*. There, both the research and publication processes rely heavily, despite a certain redundancy, on metaphors borrowed from communications and information theory. They need them, because they lack decent terminological alternatives.

EVERYTHING IS LANGUAGE

When the German weekly *Der Spiegel* decides to illustrate its investigative journalism with the reproduction of a telling detail from a state reception at the German chancellor's, it treats its readers to an inside look at the dinner menu: "Dialogue of Pike Dumplings and Crayfish on Riesling Sauce." The relation among the dish's components, the finesse of their interactions, is here imagined in linguistic terms, namely, as a dialogue. This lapse into linguistic metaphor happens not only in the case of the extravagant onomastics of haute cuisine. The hair salon tells us that it practices *Keralogie* (from the Greek *keras,* "hair," and *logos,* "learning"); the phone book lists, under foot care, a *Podologin* (from the Greek *pos,* "foot").[4]

The language of advertising is, likewise, a playground of linguistic and cognitive figuralism. The form and function of a new car are "in conversation"; a shampoo "recognizes" the needs of your hair; the color of a product "speaks to" you; the app "responds" to your movement; the fridge is "intelligent"; and everything is "sustainable" *(nachhaltig).* All the confusion of meaning in this last buzzword *(nachhaltig),* which includes the idea of something emphatic *(nachdrücklich)* in speech, of a series of consequential actions *(nachwirkend),* or indeed, as in the correct use of the word *nachhaltig,* the notion of sustainability as a quality of human planning (in which a recycling-oriented society manages to retain the same levels of material and energy, or to hold open the same range of lived and human possibility, across generations)—all this depends on, and can only be conceived through, the process of reflection in language.

The linguistic, communicative, or cognitive description of a thing seems, almost always, to involve an attempt to valorize it. When a new way of doing something enters into everyday life, we speak of "technology"; when people have mental or emotional problems, we speak of "psychology"; in both cases the connection with the *logos* or logic is so general that we lose the original sense of the etymon. But that's language, and it would be foolish to try to pin down a single origin for this boundless process of metaphorization, to assign it to the copywriters or the scientists, to braggarts or to the more general habit of thoughtless speech. In any case, all this has to do with the more general habit of confusing a thing with language about a thing (think of Magritte's non-pipe).

So you get biologists (as teachers or scientists of *bios,* "life") talking about "*biological* evolution" when they mean the evolution of living beings and should probably talk about "*biotical* evolution" instead. One need only look seriously at theories of evolution to discover their vast diversity and to see the apparent contradictions among them. If the word "evolution" refers to an actual series of historical, biotical events, then it must find itself supported by a number of competing bio-*logical* narratives.

Psychologists do no better. Though most people can tell the difference between someone who suffers from excessive fear, and has a *psychical* problem, and a scientist who is trying to come up with a rigorous definition of excessive fear, and therefore has a *psychological* problem, we have all gotten used to using the latter term to refer to the former situation.

When archaeologists explore *archaic* buildings like the Egyptian pyramids, statements about those pyramids will be true, false, unclear, or undecidable only within the remit of *archaeological* arguments. The pyramids themselves are neither true nor false but of stone.

The traditional sciences share with these examples a similarly loose and undifferentiated use of language. Physicists in the subfield of the astronomy of planetary systems forget that the objects of their study are only "physical" in the sense of being natural; they are spoken of, and researched, in the language of "physics." Objects that rotate around a central star and are held in its gravitational field do not systematize themselves but become a system only because a physicist poses questions about them or adopts a point of view toward them (about the theory of gravity, for example), approaching these natural objects through speech and action. But in everyday usage, which produces for every scientist the capacity to understand her research and results as true (and without reasonable alternatives), she treats the object of analysis as though it were identical with its representation. That the same object could be otherwise represented, or understood via a different set of statements or descriptions (and would be, by unsympathetic competitors), plays no role in the process. And so we begin to glimpse some of the motives for the tendency to metaphorize.

(It should not go unremarked that among the sciences, one major field has avoided the linguistic admixture of natural facts and their scientific descriptions, namely, chemistry. Chemistry is the only science that is not characterized by two different terms; *chemistry* can mean both the world of matter and its properties and also the world of research and theories about matter, the world of the activities of chemists. In the first case, the chemistry of things, we're talking about something as old as the universe; in the second, the chemistry of scientific research into things, we're talking about either human mastery of the properties of things that begins in antiquity or the conceptual and theoretical mastery of the properties of things as understood by modern science, that is, about either a few thousand years or a few hundred.)

Even the philosophy of science, in which reflection on language and on scientific method has played a prominent role, is not immune to confusion and linguistic sprawl. The achievements of the "linguistic turn" in the work of philosophers and thinkers like Gottlob Frege, Ludwig Wittgenstein, Bertrand Russell, Rudolf Carnap, and others seems to have been forgotten (again) today. When contemporary philosophers of science address the mind–body problem and "mental causation" (for instance, when a person intentionally raises an arm), they begin with one certainty: *physics is causally self-contained.* All of physics? Does that mean that physicists discuss only causal explanations and pay no attention to ghosts? Or is it that the subject matter of their discussion is causally self-contained? Does that mean that bodies move only by virtue of forces and not because of intentions, spirits, enigmas, and other mysteries?

Against this terminological generosity, then, we must take care to distinguish things from their descriptions—not least because descriptive statements tend to produce attributive ones. This means—to return now to communication and information theory—that whether a natural or artificial occurrence *is* a message seems more easily to depend on its *attribution* as such by an outside observer than on its representation in a controllable *description* that has to justify its own reasoning. But even this leads us to the open-ended question of whether

the linguistic metaphorization of nonlinguistic objects is only a rhetorical mode for producing intellectual and social value (in attributive
statements) or whether that mode conceals an even bigger mistake (if
the claims are descriptions), namely, the mistake of imagining that
linguistic objects have (or at least draw on) meaning and value, while
nonlinguistic objects (like the stones in the pyramid) do not. To figure
that out, we need to look at some examples from scientific disciplines
in which the terminology of information and communications theory
is standardized and common.

THE RECOGNITION OF MOLECULES

If you were to hear, today, a debate among chemists, nanoresearchers,
or molecular biologists, or if you were to glance at their scientific pub-
lications, you would discover there a highly cognitive language. "Molecular recognition" is, for instance, an expression common enough
that experts in the study of tiny building blocks no longer notice it.
Let me pull one instance of this frequent speech pattern out from
the broader whole: "The phenomenon of molecular recognition has
evolved around the principles of self-organization, self-assembly, and
self-synthesis."[5] Here molecules are granted the capacity to recognize,
which leads to an anthropomorphization of particles: in gases or flu-
ids, molecules "encounter" one another and, in so doing, "recognize"
(erkennen) whether the other is a suitable partner for a chemical reaction, so that (in the affirmative case) they can join together to create
one or more new molecules. I'm reminded of the peculiarities of the
German translation of the Old Testament of the Bible, in which Adam
and Eve "know" *(erkannten)* one another.

Because the two fields share an investment in the alleged value of
the known or the communicated, the cognitive metaphorism typical
of molecular science is tightly bound to the field of communications
technology. The reproduction of structural characteristics is consistently understood as the result of cognitive, linguistic, or paralinguis-
tic processes, as when, for instance, molecular biology speaks of the
"translation" or the "encoding" of molecular structures. We are not
talking here, then, about the journalistic popularization of highly
complex and difficult concepts for nonscientists, or of teacherly explanations by experts for laypeople, but of the disciplinary language of
the experts themselves.

This metaphorization has two major weaknesses: it is, first of all, *redundant,* relative to a causal description of the situation; and it is, second, *false,* in the sense that it involves the selection of an inadequate metaphor. To address the second weakness first: using cognitive and linguistic metaphors is unreasonable when such metaphors apply to objects or facts to which the concepts of error or falseness do not apply. It only makes sense to talk about cognition, recognition, perception, language, and other fields of difference developed through human activity when the possibility of error or of making false statements inheres in a given situation. Across all the differences of various theories of truth and value, all the different explanations for the activity of perception, the only things that can be true are also the ones that can be false: statements, assertions, and even human thoughts, where they are linguistically expressed and thereby take part in a discourse on truth or falsehood. All perceptions require the possibility of deception or misperception; they must express themselves in judgments that can be right or wrong.

The translation of cognitive metaphors into a selection process governing the modeling of molecules and chemical reactions suggests that such a process is mechanically and causally clear, functioning something like a system of keys and locks. In fact, that example can carry us a bit further: imagine that a hotel has installed a perfectly functioning locking system for its guest rooms. Every guest gets a key that only functions for his or her room, while the housekeepers have keys that work for entire floors or even the whole hotel. The system allows further differentiations of any kind, permitting, for instance, each guest key also to open the main hotel door, and so on.

At this point we need to consider two disruptions to the system, both of which seem to be based on "errors": either the guest mistakes her room for another's and finds that her key won't open the door, or she somehow—perhaps after setting it down on the buffet table at breakfast next to someone else's—ends up with another person's room key; she stands before the door to the correct room but cannot get in because she has the wrong key. In both cases it makes sense to *ascribe the error to the guest herself.* In the first case it's the selection of the lock (she stands in front of the wrong room); in the second it's the selection of the key (which opens another room, but not hers). In either situation the relationship between the lock and its key is completely unchanged; every guest key still opens only one lock. If it's tried in the wrong lock, it won't open the door. And opening is a simply mechanical, causally clarifiable matter of fact. Keys don't make mistakes; they don't "recognize" locks. Just like molecules.

The fact that these relationships can be put down in clear and ordinary sentences—like in the two sentences "the key K opens lock L" or "the key K does not open lock L"—does not permit us to conclude that their author or origin is a key or a lock itself. They originate rather in the person who discusses the situation and who checks whether the lock fits the key, or vice versa. It's indisputable that the alternatives of locking and unlocking can be decided on a purely mechanical (causal) basis. In other words, the words *lock* and *unlock* describe facts that are exhausted by the causal operations of the lock–key function. If you want to talk about "functions" more generally, you would have to bring in the goal-setting activities of the designers, manufacturers, distributors, and users of locks and keys. Only the purposive and rational actions of people in these roles turn simple causal relations into "functions," in the sense of goal-oriented procedures. And the mistakes of hotel guests who stand in front of the wrong door, or use the wrong key, are only failures to achieve their goal of getting into their own room as a result of choosing the wrong means. They have nothing to do with the causal relations of locks and keys (or of interacting molecules, for that matter).

This example, though seemingly trivial, matters more philosophically when it comes to the question of whether molecules mutually recognize each other analogously to the success or failure of a lock or key. In chemistry, too, we can find relationships that can be described and explained causally, even if mechanics does not suffice here and we have to turn to quantum mechanics.

Obviously, then, the field can include no real errors, no more than could the regime of locks and keys. Were a guest's room key to open not only her door but another one as well (against the wishes of the hotel management), the *defect* would be in the lock, in the key, or in both. Defects measure failures of human goals, not fractures in the causal structure governing the interaction of locks and keys. In this situation the causes of the defect would be investigated and explained, and the truth of this causal analysis would be attested to by the successful repair of the key on its basis.

The lock–key example suggests, then, that at the level of the object, causal explanations suffice to explain any set of circumstances or conditions. Any effort to talk about these conditions metaphorically, saying that a lock "recognizes" its key, or vice versa, is essentially incommensurable with the actuality of what takes place, since this "recognition" does not, and cannot, include the possibility of making mistakes. And this impossibility makes locks and keys differ from human users and describers of keys, or of systems of keys and locks. That is why metaphorical description is redundant, just as the first objection to the cognitive metaphorization of molecules suggested. It entails nothing beyond the causal description itself. That's why the commonly heard justification for the leap into metaphor, that metaphorization more accurately describes relations among objects, is unfounded. The truth is that they do not contain any information beyond the nonmetaphorical description and explanation.

GENETIC INFORMATION

The language surrounding the discussion of biological inheritance, of "genetic information" and "genetic codes," has succeeded in becoming part of the discourse of everyday life. Where earlier one might have spoken of "hereditary dispositions," today one makes do with the opposition between education and environment, caught between the purifying millstones of psychological or sociopolitical argument. Both parties have had their guns in position for a while, astonishing nonscientists with tales of the remarkably similar fates of twins separated at birth, on one hand, or with stories of the amazing transformations wrought by a good upbringing or good schools, on the other. But many debates in this field end up confronting the limits we all learn from daily experience: even the very best training in music cannot make a completely unmusical person into a great soloist or composer, and even the greatest natural talent cannot succeed, without the mighty striving necessary to redeem its promise.[6] Only the scientific perspective on inheritable characteristics seems to provide a reliable pathway for future research. The dissemination of genetically determined information constitutes the basis for a comprehensive, solid, and proven knowledge.

It may once again seem to be the case that either everyday language or the mediation of science journalists has ended up giving us a metaphorical way of speaking about these things. Reporting on Venter and Clinton's announcement of the successful "unlocking" of the human genome, the *Frankfurter Allgemeine Zeitung* referred (on July 27, 2000) to "the book of life," called genes or base pairs the "alphabet of the unlocked genome," and wondered if the "grammar of biology" would entail the future "readability of the world." This surfeit of metaphors belongs part and parcel to the more general public discussion of genetics, and especially of human genetics.

But the use of communications or information-theoretical metaphors appertains primarily to the *scientific disciplines* themselves. In the context of molecular biology, there are no linguistic alternatives; the metaphors are, for all intents and purposes, the primary ground of meaning. Let's look at some examples from a German textbook on genetics.[7] The "primary tenet" of genetics, Monroe Strickberger tells us, is "that this process of *information-transmission* only operates from the nucleic acid to proteins, and not in the other direction." And "the primary tenet . . . states essentially that *sequence-information,* when it has passed into a protein, cannot come back out." This tenet grounds the scientific conviction "that the three-dimensional structure of a protein—and thus its form, design, and the function that derives from them—will essentially be determined by the linear sequence of amino acids from which it is constructed. . . . In short, genetic material produces, from the *transcription* of a molecule, a *messenger-RNA* (mRNA), that complements base-for-base the bases of one of its strands. With the participation of ribosomes . . . the base-sequence of mRNA will be *translated* into an amino acid sequence" (57, emphases added).

Strickberger goes on to say that

the *transmission of information* from DNA sequences to proteins is essentially divided into two intermediate steps:

1. A *transcription process,* in which the mRNA is *copied* or *translated* from a DNA strand via the enzymatic activity of a DNA-dependent RNA polymerase.

2. A *translation process,* in which a particular nucleotide sequence from a ribosome-bound mRNA is *carried over [übersetzt]* or *translated [translatiert]* into a particular sequence of amino acids. (57)

(By the way, *translation*, in German as in English, comes etymologically from the perfect form of the Latin verb *transferre*, so the verb ought not to be *translatiert* but *transferiert*.)

These quotations show that the technical language molecular biology uses to describe and explain the fusion of an ovum and a sperm does not rely exclusively on the vocabulary of communications; we also see plenty of words whose original meanings come from human activities associated with the use of language, including translating, copying, transcribing, and so on.

So the question is whether biology's metaphorical language is primarily (1) *redundant*, because this imagery communicates nothing about the biochemical facts and their causal relations, or (2) *inadequate*, because it fails to communicate a decisive characteristic of the model it describes.

The answer cannot be different from the one for the phrase "molecular recognition." When you think of transcription or copying, you might imagine the cloistered monastic copyists of the Middle Ages, who reproduced Latin and Greek texts without knowing any Greek or Latin themselves. All it takes to be competent at such a task is to correctly redraw the marks present in the original, to "copy" them. Whether these marks are meaningful "signs," mere ornaments, or illustrations plays no role for the copyist who does not understand them. Even if he makes a mistake (leaving out a letter, a word, or even a line, transposing or doubling a letter, and so on), the copyist cannot know whether the error changes the meaning of the word, the sentence, or the entire text, whether it makes something true into something false, or vice versa. For that you need a competent speaker of Latin or Greek.

This means that words like *transcribing* and *copying* concern *two different aspects* of the production of texts, which we can compare to one another: the first is the (formal) consistency of the rows of characters, the second the consistency of the text (and its meaning). Judgments about whether a copying or transcription error has consequences for textual meaning can only be made by proficient readers. (The same applies to transcription in the sense of a shift from one font to another, for example, in the transcription of handwritten texts into machine form, as well as to the translation of one language into another.)

This second level of the understanding of textual meaning does not come into the realm of molecular biology at all. Just as we saw earlier in the example of the locks and keys, molecules do not make mistakes, which means that they cannot "recognize" anything; for the same reasons, molecular chains that pass on structures can only (causally) work or not work. Where "copying errors" appear, they do so in contexts whose explanations can only be (causally) described or explained. No higher authority judges whether any given error is meaningful. (So also for Manfred Eigen's theory of molecular evolution, in which "mistakes" in the transfer of molecular structures, modeled on the relation between a metal die and the coins it stamps, come to represent by analogy the mutation of genetic material in a given individual; this is then combined with a selection-process analogy. See later.) In short, the scientific use of metaphors borrowed from the realm of human affairs is not appropriate here, because these metaphors depend on the understanding of meaning and of linguistic value. The only metaphor that covers the situation would be something like the mindless reproduction of patterns done by a scanner or a photocopying machine.

Research into the human genome forms only *part* of the more general use of the language of communication that takes place across the biological sciences. The comprehensiveness of the context that brings together information processing and life in the natural sciences may be evaluated by the following example.

The tenth chapter of Ebeling and Feistel's *Physics of Self-Organization and Evolution* includes a section on "Storage and Processing of Information." It reads, in part,

> The ability to *store and process information* is a critical capacity of living systems. The course of biological evolution has seen this capacity go through a tremendous quantitative and qualitative growth. Three forms of information use have been primary:
>
> 1. Genetic information;
> 2. Information in the nervous system and the brain;
> 3. Extra-somatic information, stored outside the organism in records (books, etc.).
>
> These three forms have evolutionarily built themselves on top of one another.[8]

Notice the homophonic repetition of "information" in 1 and 3, which equates molecules and books! The text goes on to discuss genetic information at more length:

> The discovery of the molecular basis of, and the *principle of the encryption of information in, the genome* can be considered as one of the most important results of the last fifty years of biological science. . . . *Information-storage, -processing, and -accumulation* can be thought of as characteristics of living beings. Such beings and the structures they create . . . are complicated to such an extraordinary degree, that their emergence, maintenance, and ongoing development are essentially unthinkable without the storage and transmission of information (in experience, construction plans, or memory). On the other hand we know of *no information-processing systems* (aside from trivial examples) that exist *outside and independently from life.* (304)

Though discussing a book on the "physics" of structural organization and evolution may seem to take us a bit far afield (we are, after all, talking about molecular biology), the connection between these two fields can be found in important work done a decade earlier by Manfred Eigen, one of the founding fathers of information theory in genetics. Eigen's goal in his 1973 essay "The Origin of Biological Information" was to return biology quite explicitly to physics, aiming to *explain via physics the origin of life.*[9]

As for Ebeling and Feistel, they do not only seek a physical foundation for the biological sciences. They also discuss research in *psychology,* in which one finds the same ways of speaking. Referring to a textbook that influenced an entire generation of German-speaking psychologists (Friedhart Klix's *Information and Behavior*), they write,

> The quantitative assessment of *messages stored and transmitted by biopolymers* is extremely difficult. You must first assume that the exchange of information is always a reciprocal relationship between sender and receiver, and includes the following dimensions (Klix, 1976):
>
> 1. A quantitative and structural side. . . . This is the metrical and syntactic aspect of information;

2. A content-oriented side, that is, whatever the incoming symbols mean for the receiving system. . . . This is the semantic aspect of information;

3. An evaluative side, that is, the use and the significance, of whatever was received. . . . This is the pragmatic aspect.[10]

Here we find once again the well-known instrumentarium of Shannon and Weaver's "Mathematical Theory of Communication," familiar to us from its influence on information-theoretical language in chemistry as well as on the important language-philosophical distinctions in the tradition and understanding of Charles Morris.

At best, the adaptation of a theory developed with entirely technical goals in mind (namely, the measurement of the communications capacity of transmission channels) by the disciplines of the natural sciences is made plausible by a series of similarities:

- As it does with technical transmission, this language separates directed macromolecular processes into sender and receiver, producing a relationship between structure and process that does not depend on the means of communication, that is, on the *medium* of signal transmission. The same process happens when we talk of "encoding," which describes the technical transformation of, for instance, the sound waves produced by spoken language into electrical impulses analogized as "voice transfers" via a telephone wire, or when we speak of the technical transformation of a calculator's input into binary signs and command sequences (and then back into the sound waves of spoken language, or to the display of a decimal corresponding to the calculated result). All these resemble the language of the transfer of structural characteristics in chemical reactions.

- When the transfer of structural features of large molecules are to be described in clear language, chemists and molecular biologists use a vocabulary of verbs that refer in their original meaning to human actions, actions that, what's more, are performed in the realm of human communication and cognition.

- Borrowing language from human activities like copying or translating, or even coding or decoding (think of a telegraph operator who uses Morse code), and from the human errors

that belong to those activities (copying, translating, or transmitting something false) produces another reason to translate
information-theoretical language into molecular–biological
descriptions of genetics: in Eigen's "theory of hypercycles,"
for example, certain "errors" in the process of copying and
transcription play an important role, since they (like the
mutations in Darwinian evolution theory, almost like tiny
hereditary defects) designate the disturbances that belong by
definition to the formation and stability of a quasi-species of
RNA molecules. (This is an attempt to translate the biological concept of living beings to molecules.) In other words,
the core of Eigen's theory, in using translation to theorize the
evolution of organisms at the molecular level (including the
representation of family trees and the language of genotype
and phenotype), allows him to imagine that these processes
also involve *mistakes* in translation or transcription, analogous
to the misunderstandings and errors that take place in the
domain of human communication.

In every case, these analogistic representations forget, however, that
using these distinctions, developed in theories of technical communication and information, in the fields of chemistry and molecular biology, does not change the fact that the words in question have—both
before and outside these disciplines—meanings that are both familiar and
widely understood, meanings that they not only have but *must retain,*
so that they can function as descriptors for the modeling of molecules.
Such meanings can be explained prior or parallel to any logical, high-
minded definitional effort, with reference to examples from quotidian life. And it's precisely because of their suggestive references to
something familiar, namely, the forms of human communication in
direct speech, or the use of technical media like writing, telephones,
sound waves, magnetic tape, and dictionaries, as well as concepts like
copying or translation, that these words acquire the capacity to represent complex processes—like the transfer of molecular structures—in
theories belonging to biology or chemistry.

This kind of suggestive reference is, of course, commonly used in
model building. The model must draw from the field of the already
known (or at least better known) to represent via its modeling what

someone does not yet know (or knows poorly). But the model itself is never revised. Its dependence on the already known is methodical and thus irreversible.

One major historical and systematic foundation for this major error in thought (the idea that molecules can err) comes from the conflation of "information" with the "imprinting of form," in the sense developed by Konrad Lorenz, now classic in the biological sciences.[11] Here it is a question not only of metabolic processes in which, on one hand, the organism assimilates some set of nutrients and, on the other, uses the energy they provide to sustain itself and "adapt itself to nature" but also of "information"—initially in the scholastically derived sense of "in-forming" or "imprinting of form."[12] Lorenz also notes that, according to his ideas, processes of evolutionary adaptation "inform" organic life-forms in given environmental conditions, so that water or the laws of hydrodynamics find themselves expressed in the structure (the imprinted "form") of the fin of a fish.

This view has been strongly criticized by scholars working in the field of construction morphology, but that does not concern us here.[13] I've mentioned it only to demonstrate that in "macroscopic" approaches to theories of evolutionary or organismic development, the idea of "information" as an aspect of the three-dimensional or formal shaping of organisms or organs is an uncritically adopted and widespread principle, even in high school biology.

The reference to the idea of "imprinting of a spatial form" also appears in genetics (not least in figurative representations that show bases as differently shaped plugs and sockets and associate them with each other in pairs to illustrate the lock-and-key model). "Information" is understood as the transmission, reproduction, or copying of the spatial structures of molecules. This model, in which molecular chains serve as "matrices" for the production of complementary molecular sequences, allows the illustration of the formation of molecular chains, describable causally via chemical means, to follow the model of the metal die and stamped coinage. A spatially configured system imprints (or stamps) its structure onto another system, and this structure can, incidentally, pass its structure on in the same way, just as one can make a cast of a stamped coin. (Linus Pauling and Max Delbrück use the image of the stamp and coins in a 1940 essay on the nature of forces between molecules in living cells.[14])

So we have to ask, in what sense and with what intellectual justification can the transmission of physical structures via a casting or stamping model be identified as a transfer of "information"? What does it mean to say that information is "encrypted" or "encoded" in the physical structure of a material system? (What does it mean, if we think of the model, to say that the stamp passes its information to the coin? What information is "encrypted" or "coded" into the stamp?)

Here we get two related descriptions of what appears to be the same object: the spatial arrangement of parts in molecules or molecular chains, on one side, and encoded information, on the other. This recalls the telecommunicative relationship between signal(-structure) and message.

At this point, thanks to the daily use of information-bearing, information-processing, or information-storing machines, we are quite familiar with the material structures and changes over time that are ascribed to the transmission, transformation, or retention of information or messages. In those cases, one can say, for example, that the (geometric) shape of the grooves in a record "encodes" or bears in "encrypted" form the "information" that will be heard as music or speech once the record is played. This is a normal, customary mode of speech, for records as for computing machines, for expert systems and for machines with "artificial intelligence."

But we seem to be coming up against an epistemological problem. The relationship between the two descriptions is as muddled as it is controversial. One position, an optimistic one, claims (on the basis of one of the various forms of reductionism or emergentism) that the secondary, informational description, the meaning of the words encoded in the record, can be derived from the first, materialistic one, namely, the shape of the record grooves (Morris and Shannon both belong to this camp). Another position, more pessimistic, finds that set of claims patently absurd. But both positions need to recognize that no one has ever found a way to move from the description of material structures to the information they encode (in that direction, and in the sense of semantic and pragmatic spoken discourse, logically and definitionally derived from the shape of the grooves in the record).

Let me explain this difficulty by clarifying a few basic things, by means of a simple example. Think of a pocket calculator, the kind everyone knows. A single reference object, an individual, concrete machine, it can be comprehended via two different descriptions, or discourses of description. The first belongs to the language of engineering, of physics, which the inventor or producer of the machine uses, the second to the mathematical language of calculating processes, which mediates, as an input–output relation, between a given arithmetic calculation and its readable result. It is clear—though often sadly overlooked—that in both descriptions we are dealing with *true or valid sentences,* not false ones. The relation between the two sets of descriptions of the calculator leads to a question operating at a higher conceptual level, of how the truth of the calculated result can follow from the truth of the calculator's physics. This confronts us with a version of the famous mind–body problem: in what way do the mathematical descriptions of calculating processes depend on the technical and physical descriptions of the machine as such?

You cannot answer this question if you restrict yourself to the dogmatisms of logical empiricism and the Morrisian tradition, because you will limit the scientific character of statements either to their status as logical, definitional, and mathematical predicates or to their empirical and natural–scientific ones. The valid results the calculator produces do not *follow logically* from its technico-physical description. They also cannot be *empirically or causally understood* on its basis.

Arguments to the contrary appear in countless programmatic statements or professions of faith by optimists of neuroscience or artificial intelligence. They are easy enough to undermine. Consider the case of an error in the physical components of the calculator, whether as the result of a defect in construction or because of the sudden failure of one of its components. Either way, the failure is such that the calculator gives *false* results. (Imagine that something happens to the display, so that an 8 appears as a 0.) It does not follow, from the falsehood of the calculated result, that a technico-physical description of the calculator's inner workings is also false (as it would have to, according to the strictly logical counterargument). On the contrary, understanding the defect in the calculator, any attempt to repair it would show, begins with a *valid* technico-physical description of how it works, which would include, obviously, a description of the defect. In other words, the technico-physical description is valid *in both cases,* both for the correctly functioning calculator and for its broken counterpart. You cannot reach logical conclusions about the truth of the calculator's results on the basis of the validity of its technico-physical description.

The same goes for causal explanations, more or less. Accurate descriptions work just as well for the functional calculator as for the one that gives false responses. Giving false responses does not somehow take the calculator out of technical–scientific discourse, removing it from the general regime of statements that provide valid causal explanation. The calculator's physics does not causally explain the arithmetic of its performance. Naturalists and logical empiricists have simply forgotten that we are not dealing with just any physical and mathematical descriptions of the pocket calculator but with *valid, true* ones. No one cares about a calculator that produces strings of arbitrary signs. People want a calculator that gives the *right answers,* which is why they invent, build, and use calculators in the first place.

The insights from this example transfer to our more immediate field of concern here, namely, molecular biologists' reliance on words like *transcribe* or *code,* which color their language with information theory. No road leads from the description of material systems *to the validity* of descriptions of so-called higher system-characteristics. The metaphors used in genetics today are thus not only redundant but inadequate to their scientific tasks.

FROM METAPHOR TO METHODICAL HEADSTAND

In the natural sciences, using metaphors to model processes has any number of advantages. As long as the metaphorization of a state of affairs does not directly contradict its basic properties, the choice of a metaphor will be appropriate; there's really nothing inherently wrong with the scientific use of anthropomorphic metaphors. In fact, anthropomorphizing machines is a normal, unproblematic aspect of everyday life: "the car doesn't want to start"; "this computer hates me"; "the TV remote has wandered away again." These conventional ways of speaking "as if" can always be easily resolved, since the technical–artificial character of the object is clear, and, at worst, the producer, seller, or user of the machine is actually to blame for the problem.

Things are otherwise when, in the course of exercising their disciplinarily specific competence, scientists treat the anthropomorphic qualities of their information-theoretical or communications-oriented modes of speech as though they were *not metaphors at all* but rather actual, direct, original, nonmetaphorical representations of scientific processes. This shift, this slippage, lies at the very heart of the naturalization of information.

If we follow the earlier citations from the world of biophysics, then information (and its creation, processing, and storage) is a specific property of living beings. Were that true, obviously, the fields that study life, the biological disciplines, would be responsible for its study. Should it then be the case—as in fact it actually is today—that these sciences take as one of their projects to explain *the historical emergence of life from nonliving substance,* biology would then be bound also to attempt to understand the *origin of information itself.* Finally, because physicists and chemists are responsible for understanding the conditions and development of the universe during those times, or in those spaces, in which no life exists, then the "biological" origin of life, and information along with it, would have to be handled using the methods belonging first and foremost to those fields. In the final instance, then, physics would be, if only it were finalized, the science responsible for describing the history of the universe from the big bang to the social welfare state, including, of course, the history of culture, the history of the natural sciences, and, finally, the ideas considered in this book.

A thoughtful person can, one assumes, recognize in the universalizing extension of such a program the risks of epistemological overboldness and the many problems of circular reasoning. It would therefore be a bit surprising if anyone seriously took such a position.

And yet.

Consider, for instance, the book *Consilience: The Unity of Knowledge,* written by the American biologist Edward O. Wilson, whose public profile makes him an exemplary case here. Wilson writes, "The central idea of the consilience world view is that all tangible phenomena, from the birth of stars to the workings of social institutions, are based on material processes that are ultimately reducible, however long and tortuous the sequences, to the laws of physics."[15] What Wilson refers to as a "central idea" and a "worldview" occurs, however, not as a bold thought experiment but as a statement of fact: "It is the custom of scholars when addressing behavior and culture to speak variously of anthropological explanations, psychological explanations, biological explanations, and other explanations appropriate to the perspectives of individual disciplines. I have argued that there is intrinsically only one class of explanation. It traverses the scales of space, time, and complexity to unite the disparate facts of the disciplines by consilience, the perception of a seamless web of cause and effect" (291).

At the foundation of Wilson's central idea lies the explanation of cause and effect in the sense of natural–scientific causal knowledge. Anyone who would want to make a claim about culture, language, or the social sphere, in order to recall the potential responsibility of actual, acting human beings, will come across these defensive statements at the beginning of the book: "Given that human action comprises events of physical causation, why should the social sciences and humanities be impervious to consilience with the natural sciences?" (11). The philosopher, gasping for breath, keeps reading: "Philosophy, the contemplation of the unknown, is a shrinking dominion. We have the common goal of turning as much philosophy as possible into science" (12).

This striking example of naturalization has a number of predecessors in biology. Konrad Lorenz, who, like Wilson, won the Nobel Prize, attempts in *Behind the Mirror* to "search for a natural history of human knowledge" in an epistemological sense, aiming to produce nothing less than a full naturalization of the work of his Königsbergian predecessor Immanuel Kant.[16] Where Kant speaks of the conditions of possibility of experience, in Lorenz the *a priori of knowledge becomes the a posteriori of phylogeny*. The knowledge-dispositions (*Erkenntnisdispositionen*) that Kant imagined are, in such a model, the natural–historical result of evolutionary selection. The expansion of evolutionary biology into the philosophy of knowledge serves not only to naturalize human cognition but to bring the entire history of human culture under the explanatory regime of the natural sciences, from whose perspective culture—singing, books, whatever—looks like nothing more than a succession of natural events.

The modesty of Lorenz's tone has nothing, however, on the boldness of his arguments:

> There are in my view definite signs that a self-recognition
> of all cultural humanity, a collective self-knowledge derived
> from natural science, is beginning to spring up. . . . *A reflecting*

self-investigation of human culture has never yet come into being
on this planet. . . . The scientific investigation of the structure
of human society and its intellectual processes is a task of
mammoth proportions. . . . Yet I believe that the human species
stands at a turning point in history, and has at this point the
potential capacity to scale new and unknown heights of its
humanity. . . . The modes of thought that belong to the realm of
natural science have given us the power to ward off the forces that
have destroyed all earlier civilizations. This is so for the first time
in the history of the world.[17]

What more is there to say?

Let me be clear: I am not interested in criticizing Lorenz's ambition (or his hubris); I am tracing the mythology of information's naturalization. More specifically, I am concerned with the ways in which
metaphors drawn from linguistic and communications theory become, via reinterpretation, descriptions of scientific objects.

The naturalizing strategy cannot deny that language, the philosophy of language, and semiotic theory (at least in part); information and
communication technology; and indeed the entire system of differentiation and hypothesis making are all historically much older than the
radical naturalization of information by the younger sciences of biology and physics. On the contrary. Naturalizing thinkers essentially
believe that the humanities and philosophy are older, less developed
predecessors of better, natural–scientific insights. Their story goes
thus: first we had myths, then we had philosophical explanations of
myths, then the emancipation of the sciences (*Teilwissenschaften,* the
"branches of science") from philosophy, and finally the rectification
of the philosophy of the Enlightenment and the humanities by the
natural sciences. That's the naturalizer's history of epistemological
progress.

For those reasons, naturalizing thinkers need not deny that "first"
came human language, then the (still linguistic) reflection on language, so that there was something for the natural sciences to over-
come, to naturalize, in the first place. The downright eschatologi-
cal end of this developmental pathway aims to attribute absolutely
everything to physics, with the particular goal (not polemically implicit but expressly articulated) of turning a history of physics and

biology (considered as part of the history of culture) into a physics of history itself.

We're still in the early steps of this transformation: "When we describe the principle of molecular information-storage in the genome of a living being as a 'molecular language,' we are speaking of more than a simple metaphor." So claims Bernd-Olaf Küppers (a student of Eigen's) in *The Origin of Biological Information: On the Natural Philosophy of the Origin of Life*.[18] He justifies the assertion: "This is particularly clear in the syntax, in the hierarchical relationship of individual symbols with each other. Genetic language has, just like human language, a syntactical dimension." Küppers follows that sentence with a table that translates "genetic script units" into their "analogical language units." Nucleotides correspond to letters, codons to phonemes or morphemes, genes to words or (simpler) sentences, scriptons to (compound) sentences, chromosomes to paragraphs, genomes to complete texts, and genotypes to complete texts with commentary.

This leads him to conclude that "the *analogy* between human language and the language of genetic molecules is *completely stringent*." Although, apparently, "the language analogy . . . has its limits. If one disregards certain phenomena of molecular regulation, the language of genetic molecules has no interrogative features. And the generative character of human language finds its molecular-genetic equivalent only in the area of the evolutionary development of living beings as a whole." Nonetheless, Küppers suggests, with the help of the biologist's suitable understanding of the Shannon–Weaver theory, and a correspondingly serious interpretation of semantics and pragmatics, it should be possible to develop a semantics of genetic information and, with it, the capacity to interpret genetic information in pragmatic terms.

You can't be too philosophically meticulous when reading a text like this. The relation between human and molecular language constitutes a "completely stringent analogy"; a few sentences earlier we were told that we are dealing with "more than a simple metaphor." Metaphors describe movements of meanings between two fields of understanding; analogies describe structural or functional similarities. To be clear, Küppers would have to decide which of those two concepts he was using and would then have to tell us whether he means for the genome/language analogy to indicate a structural or functional resemblance. That decision would allow us to understand whether we were dealing here with an essentially arbitrary arrangement of molecular and linguistic building blocks or with a language capable of bearing meaning and value!

As for the idea that "certain phenomena of molecular regulation" may contain "interrogative features," it suggests that we ought to interpret a comparison of actual and desired values in the control process of "genetic molecular language" as a question. Like the program of the book as a whole, this moment aims to reduce the foundations of higher, cultural phenomena like human language of the history of art to a set of physical descriptions and an explanation of the "origin of biological information."

I don't want to spend too much time on the scientifically dubious gap between claim and reality, between the program and its carrying out, that characterizes this work. There's not much going on here other than a highly speculative (structural) analogy between human language broken down into component parts, on one hand, and a somewhat violently reimagined version of molecular "language," on the other, even if we are told, on the basis of some kind of gut feeling, that the analogy is "stringent." There is simply no evidence that molecular processes resemble the use of language as a means of communication between rationally acting human beings.

What one has every right to expect, however, from an author who deals with philosophical topics is the very basic thought that *his entire program,* and indeed the execution of the book he has written, *is always already taking place in language.* Küppers's arguments themselves rely on appeals to the concepts of understandability and of internal coherence. He overlooks this basic fact. But the naturalization of language via the building of analogies cannot build the ground under its own feet; because it takes place in language, it constitutes a *performative contradiction.* Molecules and their structures do not ask questions or make claims; they do not make statements about their inner states or perform speech acts like requesting, thanking, promising, or deputizing. No attempt to redescribe molecular biology in information-theoretical terms can therefore succeed. No road leads from what molecules actually do to the world of speech.

In other words, we have here transgressed the basic principle that the natural sciences of human beings should not describe these latter in such a way as to exclude the description (and indeed the act of describing) from itself. Let's call this the *culturalist anthropic principle*. Such a principle has to be strictly distinguished from the naturalistic "anthropic principle" of contemporary astronomy, which constrains theories of the formation of the universe. In its weaker form the constraint states that such theories cannot rule out the emergence of human beings; the stronger form requires those theories to *explain* the origins of humans on the planet Earth. Both versions conceive things naturalistically, because in both, "humankind" emerges only as an object of biological science and thus as a natural object. This dogmatic restriction aside, one must demand from cosmology, and from the natural sciences concerned with humans more generally, that they not exclude, at least in the weaker version, the workings of their disciplinary fields. But because the claims of science about humankind aim specifically to replace humanistic or philosophical descriptions with scientific ones, *the principle must be fulfilled even in a stronger form*: the biological sciences must be able to explain the capacity of humans to have developed the biological sciences and the fact that they have actually done so.

The violation of this principle in the theory of the origin of biological information and elsewhere ends in a naturalistic short-circuit, a kind of methodical headstand, in which the explanans (the thing that explains) and the explanandum (the thing that needs to be explained) trade places. Nature no longer explains the cultural phenomena of language and science; rather, the cultural phenomenon of science explains natural phenomena. It does so thanks to a discourse and a language borrowed from the specifically cultural successes of physics, biology, and information theory.

The Spiritless Machine

From the history of misunderstandings associated with the communications complex—including the dogmatic logical empiricism invested in it, its elaboration in the chatter of molecules, and the theory of the primacy of biological information and the methodical headstand it performs—one persistent motif emerges: the conceptualization of information as merely structural or syntactic. Such a conceptualization brushes aside questions of the meaning and value of linguistic communication. It owes its authority to *a fundamental ambiguity in the realm of syntax and the description of structures*.[19]

The spatial organization of things like patterns on paper, molecules in a molecular chain, grooves in a record, or magnetized particles on a hard drive is neutral when it comes to bearing meaning and value. (The grooves on the record work just as well for value- and meaning-laden music as for a spoken lecture.) The structural theorist turns out to be naive with respect to his own value claims. He extolls them like the thoughtless layperson in the simple course of everyday life. But anyone who claims makes a claim and nothing more; no description or reflection is implied. If the claim is denied, the layperson reacts within the frame of a more or less normative dialogue. In the case of the reduction of human language to its physical medium, however, the naturalizing theorist expresses an essentially fatal naïveté. He treats the human use of communication as a means of transmitting meaning and value as identical to the work of a telephone, which follows the laws of physics. But it makes no difference to the functioning of the telephone whether what it passes on means anything, whether what it communicates is true or false, useful or useless, understandable or not. The same goes, obviously, for the naturalizing information-concept and its scientific assumptions.

Together, this loss of the syntactical or structural aspect of the content of communication and the various elaborations of the naturalizing perspective on it expose a *mind–body problem*: how can the physical media of information or communication possess or bear meaning and value, that is, how can they have qualities that have something to do with the forms of understanding and recognition shared by participants in a communicative process? At this point the naturalization of information confronts a dilemma that is not limited to theories of information or communication. It is a side effect, sometimes a cause, sometimes an effect, of the unresolved mind–body problem, a problem that appears not only there where human science takes on, with the help of physics, the task of describing both mind and body as material systems, and hopes, in so doing, to rise from those descriptions to explanations of the higher mental qualities of human action and consciousness.

The mind–body problem is not simply an issue for human beings themselves. It includes the entire domain of human production and creation governed by the apparatuses of communication and computer technology as well as the sciences that aim to describe and explain them. That's why I discuss in what follows those aspects of the mind–body problem that are closely related to the naturalization of the information-concept. This will allow me to criticize the idea that the *reduction* of communication to syntax or to the structure of message transference frees the scientist to move back in the opposite direction, ascending from that initial reduction toward meaningful human discourse. Such "ascents" make use of their own theories; here I'll focus especially on systems theory and emergence theory. That will lead us to an understanding of where the mind–body problem, today largely discussed in neuroscientific terms, really lies.

INVENTING THE MIND–BODY PROBLEM

In the practice of ordinary life, we don't experience a mind–body problem. A person goes on a walk, eats a snack, or has a conversation with someone. For the observing scientist, these actions have both corporeal and mental components. Only in the case of particular disturbances—the walker trips on a stone, the eater bites down on a cherry pit in a cookie, or the speaker is drowned out by the supersonic blast of the Concorde overhead—will body and mind be differentiable. In the case of these corporeal disturbances, the body is like a suddenly malfunctioning instrument, interrupted in the pursuit of a particular activity. "Mental" aspects of the situation—the recreational value of the walk, the pleasure of the cookie, or the novelty of the conversation—suffer as a result of these physical annoyances. But at the same time, these annoyances bring the nonphysical dimensions of experience to our attention. In short, in ordinary life, the mind–body problem takes the form of disrupted relationships.

The question about the physical causes of disturbances of undivided processes of ordinary life emerges, not as matter of philosophical luxury, but as a helpful intermediate step that pushes us to recognize the causes of these disruptions and thus to fix or avoid them. The mind–body problem is, in other words, a useful schema in the general realm of the management, avoidance, and remediation of disturbances. Such knowledge is, in ordinary life, a matter of know-how.[20]

In the usual run of life, however, we confront less a mind–body *problem* than a mind–body *difference*. The *problem* exists in science and philosophy, when one seeks either to "explain" the mind through the body or to "return" from, or reduce, the latter to the former. Here we may well want to address the history of the mind–body, or soul–flesh, problem from a historical perspective, because the relevant literature in the (analytic) philosophy of mind dates back to antiquity and the work of Plato (a dualist) and Aristotle (a monist).

We call *dualistic* those claims that postulate (in a variety of forms) the existence of two different worlds: a physical, material one and a spiritual, immaterial one. Methodically, the attempt to understand the interrelations between the worlds only takes place after their division. Those interrelations, at least in the sense of actual interactions, are easy enough to see. If a person throws a stone, she has turned something mental, namely, her desire, into physical movement. The person who gets hit by the stone, and feels pain or maybe anger at the stone thrower, has turned a physical movement into something mental.

The familiarity of these ordinary examples only produces a mind–body problem through a certain amount of inventive work. This inventiveness has less to do, however, with the exemplary situations in which the mind–body aspects of ordinary life fall apart (at least from the observer's point of view) than with the sciences that take responsibility for the material or embodied domains of being (physics, chemistry, biology). These sciences distinguish between methodologically legitimate causal explanations and methodologically illegitimate explanations of mental states. And that's where the problem begins.

Monistic points of view recognize, on the other hand, the pseudo-problem produced by the dualistic approach. Consider, for instance, my earlier description of a pocket calculator as the product of a dualism in description but not in substance. Here you have no theory of two worlds. But when monistic approaches fall under the dogmatism of logical empiricist philosophy, and thus only recognize causal or logical explanations as scientific, monism has no chance of success.

How does it happen that an entire philosophical and scientific tradition has failed to understand the mind–body problem? You've seen part of the answer a few times already. It has to do with the (ungrounded) limitation of scientific means of investigation to the regimes of the mathematical–logical–definitional (the so-called analytic truth), on one hand, and the scientific–empirical truths developed by the thinkers of the Vienna Circle, on the other. These conditions still hold true today, even if the existence of analytically true statements has been put into doubt by W. V. O. Quine, one of the major figures of analytic philosophy.

Part of what makes the mind–body problem hard to understand is that it has opened, for the natural sciences and the philosophies associated with them, important, new, and highly attractive fields of study, which exist precisely to get around it. I'm thinking here of systems theory as well as theories of emergence and supervenience. In each of these—to generalize a bit—aspects of mental life are conceived as higher system-characteristics that come into being wherever the physical is organized in a sufficiently complex way.

Thanks to the recent popularization of neuroscience, even lay readers cannot have avoided learning about the brain's hundred billion neurons and their astonishingly high number of connections, which together produce a supercomplexity that may never be fully understood. At the same time, we attribute the brain's highest accomplishments to its character as a system, accomplishments we encounter on a daily basis in consciousness, perception, language, and so on. For the naturalization of information, this means that the brain, now conceived as a "syntactic machine," would by virtue of its supercomplexity be especially able to manage the problem of the meaning and validity of linguistic expression, that is, the semantic and pragmatic aspects of language.

Following systems theory, theories of emergence and supervenience—eagerly embraced by scientists, especially evolutionary biologists—focus on the relationships between components or compartments of entire systems. The idea that the whole is greater than the sum of its parts (fatal motto!) is happily exemplified by the relation between individual notes and the complete melody or between a set of gears and the clock they make. Only the sequence and duration of the notes make the melody; only the system of mechanical components, and not the individual gears themselves, measures time.

And so the naturalistic mind comes to wonder if humans and their culture originate, evolutionarily speaking, in the primordial soup that existed in the first days of the planet Earth and the universe, at which point it stands to reason that we could interpret these relationships as chronologically ordered steps to higher development—all this naturally taking place, of course, only under the dominion of the laws of cause and effect. The higher characteristics of such a system are called, relative to the lower ones, "emergent"; or, if we want to illuminate the descriptive quality of these relationships, we will say that the description of the higher system-characteristics is supervenient on the description of the lower ones. So much for a provisional, general introduction.

PROBLEMS OF THE NATURAL AND TECHNICAL SCIENCES IN SYSTEMS THEORY

"Systems theory" is today a well-known expression, a long-established sign for a field that has earned the right to define areas of research and application. There are introductions to systems theory, and textbooks for it, just as there are for classical mechanics or the mathematics of infinitesimals. But what is the object of systems theory? Let's take a look at the history of system-theoretical approaches in academic disciplines. I leave aside for now that "system" is also a term in philosophy, that it plays (for instance, in the work of Immanuel Kant) an important role there; neither will I account for the effects of the philosophical history of the concept of systems in the earliest claims of technical and natural science.

It's instructive, however, that the system-concept is already of great significance for the naturalization of physics. In chapter 2, I discussed Heinrich Hertz's role as a source for the naturalistic understanding of physics. The sciences owe him thanks not only for his introduction of the concept of models in physics (which then flowed to the other natural sciences) but also for an extensive use of the idea of systems. When it comes to movements and their explanations by forces, real bodies are, in Hertz, idealized as systems of mass points. System and model are bound tightly together: "A material system is said to be a dynamical model of a second system when . . ." begins the first definition in the section "Dynamical Models" (175). It ends with an "observation": "The relation of a dynamical model to the system of which it is regarded as the model is precisely the same as the relation of the images which our mind forms of things to the things themselves. . . . The agreement between mind and nature may therefore be likened to the agreement between two systems which are models of one another" (177).

"Systems" are, for Hertz, abstractions, representations, and idealizations of objects in two different areas; they have a (symmetrical) relation to one another that takes the form "x is a model of y."

The historical development of systems theory took place less around mechanics, however, and more around electricity, in which a new problem field emerged. The wide understanding of Ohm's law, which governs the relationship between voltage, amperage, and resistance, and the mastery of inductive and capacitive phenomena in electrodynamics, led to a concern with the control of electrical circuits. How do currents flow, and how are voltages distributed, in electrical networks?

By the middle of the nineteenth century, two competing ideas emerged: mesh theory and node theory. Following the general principle that electrical energy cannot come from nothing and cannot disappear into nowhere, scientists postulated that in a node at which several electrical lines meet, the incoming and outgoing voltages cancel each other out. In a (closed) mesh in a network, the same thing happens. (Both of these amount to establishing a kind of design characteristic of electrical networks. Clearly, a node can be neither a power source nor a power sink, and a mesh of electrical conductors cannot be a perpetual motion machine, producing endless power without input from an external power source.)

One important breakthrough for the contemporary understanding of systems theory came from a German electrical engineer, Karl Küpfmüller. Küpfmüller had a fantastic idea: he put together a schematic representation of an electrical network by using only two types of components, a bipole and a quadripole. A bipole is a black box with two openings, the simplest example of which is a piece of wire. All you need to know about the black box is its resistance in ohms. Lightbulbs, doorbells, refrigerators, and any manner of other electrical appliances can be described in the same way. A quadripole is, by contrast, a black box with four openings. Anyone who charges her cell phone with a power adapter is using a quadripole: the two poles of the power plug and the two poles of the phone's power supply. Transformers, amplifiers, transmitters, receivers, and many other machines are quadripoles.

Küpfmüller's idea was simply to represent any known or possible circuit only in bipoles and quadripoles. The task of the electrical engineer is, at that point, simply to reckon the components (the black boxes) according to their input and output ratings.

One feature of Küpfmüller's thought really matters for a better understanding of the concept of systems: his systems theory is a specialized means of representation. Even someone who cannot read circuit diagrams or use electrical instruments knows that the representation of a circuit diagram has a particular relation to the actual apparatus in question: the diagram represents the functional properties of the machine. It shows how the machine is built, from the perspective of its electrical operations. Circuit diagrams are necessary when a defect happens and its cause is sought in the malfunctioning of a component, so that, once it's exchanged, the machine will work again. Anyone who owns a car knows how this process works.

This form of systems theory amounts to a description of complex, coordinated objects primarily focused on their particular functions, oriented finally toward the functioning of the totality of the system as it depends on the functioning of its components (part-systems). From the simple flashlight with a battery, the lightbulb, and the switch to the far more complex workings of a digital camera, we are always speaking of artifacts, that is, of machines produced by humans for the technical–artificial fulfillment of specific goals and designed therefore with specific capabilities. Systems theory is a means for describing these capabilities and their functionality.

It's useful here to stress the emergence of systems theory within the framework of engineering, because that field, unlike biology or sociology, focuses especially on systems that are cultural, "technically" (in the Greek sense) imagined, invented, built, and produced by people in the service of their own chosen goals. Whereas certain philosophical attitudes from biologists or sociologists can be seen to imply that the systems they study (like the solar system or the social system) emerge "spontaneously" or "naturally," and are thus a priori subjects of scientific research, the engineer has no such option. Even in the special case in which an expert has to analyze the function of a totally unknown technical system—as might be the case when military scientists reverse engineer an enemy rocket—the engineer is still dealing with an object built by people according to known physical theories, that is, an artifact.

The objects studied by engineering thus permit us to observe that the *typical problems of describing system functioning*—how the parts of the system work together to create the action of the whole, and how the function of the whole determines the working of the parts— *are not the methodical beginning* of the creation of systems. In the creation of systems, things move along nicely, step by step. The person who draws a circuit diagram does so line by line. Step by step moves the person who diagrams the circuit after the fact; step by step works the producer of some new electrical or electronic machine. In none of these processes are the steps arbitrarily transposable. This is true also for chains of production, where the penalty for transposing steps is a failure in the object's desired function. Indeed these relationships resemble those in any productive chain of actions. If you want to make a painted wooden statue, first you carve the wood, and then you paint it.

It's only in the case of textbooks in systems theory, or in the application of systems theory to biological, economic, or social systems, that the authors incessantly complain that the apparently one-dimensional sequencing of words and sentences in a text cannot be an adequate representational medium for system functioning. To pick one example at random: "Some problems in systems theory (for example, border—structure—process; structure—function—system; action—anticipation—system) are so tightly interwoven and so interdependent that they must be presented simultaneously. That simultaneity is impossible in the medium of written language."[21]

We've arrived at the famous conundrum of the chicken and the egg. That the function of the complete system cannot be produced from anything but the partial functions of its components (recall the idiomatic "more than the sum of its parts") leaves us no royal road to a solution. Someone seeking to represent, describe, or explain a system would have to start with any one of its partial functions, or examine the system's interactions with its environment, then move on from any one of these to some other undescribed element.

This leads to enormously confusing representations in, for instance, the biological theory of radical constructivism, theories of society in the social sciences, or theories of complex economic systems. It also produces some specific problems for the validity of systems theory, because gradual or ordered substantiation is simply excluded from representational legitimacy.

If one considers, however, the system's character of being-produced (Hergestellt-Seins) via the origin of systems theory in engineering, one can illustrate (ironically) this difficulty in the following way. A woman looks at a carpet, indisputably hand-produced, in which wool threads have been laid down diagonally along the intersections of the warp and woof threads. To make the carpet, the threads were stretched as horizontal and vertical strings into an upright, rectangular wooden frame, and then the knots were knotted. When the carpet was ready to be taken off the frame, the snipped threads in the warp and woof had to be knotted so that they could not be pulled off the edges of the carpet. A good, durable carpet will have its knots placed densely enough to hold the warp and woof in the proper tension.

Now, someone could reason her way to the following idea: on one hand, the warp and woof threads must be tensioned so that they can bear the weight of the knots. And on the other, the knots must be packed tightly enough to help tension the threads. Which one of these is the chicken and which one the egg? The warp and woof threads are conditions for the possibility of the location of the knotted wool, and the knotted wool is the condition for the possibility of the threads. Here we arrive at the very descriptive difficulties (and their bases) that systems theory complains about: the woman is only observing the finished carpet and not the story of its making. But the chain of production steps happens in irreversible order: first the frame, then the stretching of the warp and woof threads, then the knotting of the wool.

In other words, the problem that plagues systems-theoretical descriptions in the natural and social sciences is simply this: that it would be nice to grant the system *the status of a given,* without having to think about *its genesis*—its genesis either as a system or as the description of one. In the case both of natural objects like the solar system and of cultural ones like society, it is the criteria of differentiation brought to the object by its describer that define the system as a system in the first place, or, in the case of artificial systems, that produce the carpet. And this production, like the system of descriptions, has to take place in a methodical order. Otherwise, there would be no system to look at, not even in the description!

These paradoxes of systems-theoretical analysis—including the tricky ordering problem, wherein the systems theorist can find no place for himself in the (social) system he describes, and can make no truth-claims about the value of his own description—are in fact mirages. They resolve themselves when you abandon the prejudice toward the givenness of systems and recognize, instead, their constructedness. To construct, to produce, even when construction or production is only a matter of describing something that already exists, is subject to the constraint of following an order of individual steps that lead to success. All it takes for systems theory to resolve its paradoxes and irritations is to act methodically, to consider the context of rationally ordered creation instead of just the context of the functions. Functions are not metaphysical entities but goals, established by the people who initially conceptualized the system.

Let's return now to information and its allegedly natural character. At this point it is possible to assert *for any system that works on information* (including its transmission and storage) that, insofar as it is artificial, that is, made by humans, its understanding requires the step-by-step, goal-oriented method. As we saw with the mechanization of communication (in chapter 2), the invention of the gramophone or telephone as technical means of information transmission depended on understanding, analysis, and goal setting that begins with the speaking and listening of two participants in a conversation. Only from that initial scene could the desire emerge to transmit spoken language over distances of space or time. And only on the basis of that desire could someone conceive the system whose goal would be to fulfill it. The same goes for pocket calculators and photocopiers.

It takes, therefore, no real philosophical astonishment to grasp how a gramophone speaks, how a dictation machine understands words, or how a calculator calculates. All these actions are essentially system functions that come only from the so-called mental setting of goals, which are then fulfilled by material means, namely, the machines in question. The amazing, *higher* system-characteristics are in fact the primary, desired, and perfectly understood things that the artificial system was built to do. Only the foolish reversals of reductive thought find themselves confronted with a problem, when they try to understand the higher system-characteristics from "below."

In historiography of nature, things go the same way. It's not the case that the earlier stages of an evolutionary process are the known ones and the later, more complex stages involving mentally complex humans the unknown. The history of nature as a series of natural events is often studied "backward," starting from the present as a fixed point. How can—in the case of the known present resulting from known laws—the present as product of the past be traced back to a hypothetical past? The idea that the emergence of higher human or even animal capacities begins from the activities of single-celled organisms appears only thanks to the construction of a model. Which is to say that it appears in exactly the same way that we invent, build, and use the artifacts of information technology. We only know nature through our constructions of it, all the more so when it lies in the past.

PROBLEMS OF NATURALISTIC PHILOSOPHY IN EMERGENCE THEORY

The other major concept that theorizes a transition from physical or even general scientific description of a system to its higher characteristics goes by the promising name "emergence." The name borrows its etymology from Latin, combining *ē*, out, and *mergĕre*, to dip, and for that reason also means something like "to come into view." In the history of biology it had one unsuccessful competitor. In his chapter on the evolution of the human capacity for knowledge Konrad Lorenz borrows the word *fulguration* from medieval mysticism and explains it by comparison to electrical circuits: the connection of two circuit boards produces "entirely new, unexpected system characteristics . . . of whose appearance there was previously not the slightest suggestion" (30). *Fulgration,* from the Latin *fulgur,* "lightning bolt," refers to this sudden appearance of fully novel characteristics. Lorenz continues: "Although, as we must always emphasize, we scientists cannot believe in miracles—that is, in violations of the universal laws of nature—we are at the same time aware that we can never succeed in giving a complete explanation of how a creature has evolved from its lower ancestors. A higher animal . . . cannot be 'reduced' to inorganic matter and the processes that take place within it. The same is even true of man-made machines, which provide a good illustration of the essence of this non-reducibility" (35, translation modified).

Though one cannot simply equate Lorenz's *fulguration* with the contemporary word *emergence,* his explanation of the concept nonetheless offers a concise and pointed entry into one type of emergence theory. It has to do with whether we can (or should) accept the notion of unexplained (or unexplainable) causation without letting go of the entire apparatus of the scientific explanation of causes.

We can compare this to the earlier example about systems theory: just as a melody has qualities that do not belong to its single tones, and a mechanical clock does more than do its deconstructed parts, so here, in research into the *history* of nature, we find that living beings have qualities that differ substantially from those of lower ones. More primitive creatures are naturally not components of higher ones in the same way as tone is a component of melody or a gear is a component of a clock (or, for that matter, organs are components of organisms) but in some sense are precursor constructions, more akin to the simplified versions of components an engineer might use in the preliminary stage of building a complex circuit.

Already here we can remark on a language-philosophical shortcoming of fulgration or emergence theory. It cannot be too controversial to notice that, whether we speak of tones and melodies, gears and clocks, organs and organisms, or, finally, the capacities of an amoeba versus the capacities of a falcon or a human, the only thing that is comparable is the *language of description*. At that point we would need to *justify* our process, explaining exactly *why* the language for describing the respectively higher or more complex systems should have to be "reduced" to the language for describing the systems that are relatively lower or simpler. Natural scientists who find such a claim plausible overlook the simple fact that "reducing" something is a human action and not a natural process. Actions have the peculiar quality, after all, that one can also refrain from doing them!

These category errors are only possible because scientists fail to notice that the biological and physical sciences are the products of human activity and so subject to validity claims *(Geltungsanspruch),* while the objects of these sciences, life or matter, make no claims to validity. To say it another way, *there is no reason to expect* that the language patterns used to describe various kinds of objects could be reducible to one another or be standardized in some way. The naturalistic profession of faith (even Lorenz speaks, not incidentally, of "belief") in a *single* nature obeying *one* natural law, and therefore in *one* science, speaking a *single* language, repeats a familiar error—the same one Hertz made in the naturalization of physics. With a little sensitivity to the problems of the philosophy of language, anyone can notice that a conversation about oil paintings deals with a different realm of research from a conversation about oil paints, without therefore deciding that any discussion of the meaning or value of oil paintings can only possibly take place in the language appropriate to discussing oil paints.

But here the faithful naturalist raises his hand and says, "Yes, but" . . . which is why we have to come back to emergence theory and its role in the naturalization of the information-concept. Achim

Stephan's *Emergence: From Unpredictability to Self-Organization* offers a useful overview of the current state of the field. Before getting to the book's systematization of contemporary emergence theory, I want to say two things about the way it thinks:

1. All the concepts and theories that the book presents, analyzes, and evaluates begin from an "unreserved acceptance of the naturalistic approach."[22] Stephan says he will show that "everything in the natural system is put together from the same basic building blocks," so that there are no basic substances, for instance, that living things have that nonliving things lack. These claims are followed by citations from various major figures in emergence theory, most pertinently R. W. Sellars: "It is evident that the spirit of naturalism is identical with the spirit of science."[23] Naturalism and science are, for Stephan as for Sellars, identical. This commitment to naturalism allows theorists of emergence to stake "their empirico-scientific claims" (15) while "limiting themselves . . . above all to those explanations of the world that explicitly exclude supernatural entities, whether as guiding components of organisms (entelechies) or more generally as motive factors in the evolution of natural processes *[élan vital]*" (14).

   In short, whoever wishes to follow theories of emergence enters the garden of naturalists and thereby locates the rest of the world in the terrain of the nonscientific, the obscurantist, the metaphysical, and the religious. The reader will search in vain for a grounding for or justification of this decision. The whole thing must give, for the author, that sociopsychological feeling of belonging to a community of like-minded thinkers, of the pleasure of interacting exclusively with one's peers and their theories. The question of whether the *foundational decisions of naturalism* themselves are *reasonable,* or could even be formulated clearly without these quasi-religious acts of faith, lies outside the naturalist garden.

2. The second surprising thing in Stephan's book lies in the manner in which the author handles his own questions. The book neither describes nor explains emergent phenomena (the language of the sciences), nor does it discuss the concep-

tual or methodological means used by natural science (the language of scientific metadiscourse). You do find a few famous examples, like the one of how carbon atoms organized in the form of graphite or diamond demonstrate two extremely different sets of higher system-characteristics, which ostensibly produces an emergence problem for chemistry. But these examples are barely discussed, and no one ever asks with what means a chemist arrives at the concept of carbon or with which methods she decides that graphite and diamond are forms of its appearance.

The object of the book is thus neither emergent phenomena nor their scientific description but emergence *theories,* which talk *about* scientific problems and thus generate a third level of discourse. Talking about these theories generates, then, a fourth level of discourse, namely, a metadiscourse involving the representation and discussion of the *competition among various theories of emergence.* This discourse operates necessarily as a kind of philology and history of emergence theories themselves.

The reader must therefore become familiar with the style of analytic philosophy that considers only texts as the objects of investigation and limits the selection of those texts entirely to their own philosophical circle and its history. The book gives no signs that its author knows about the level at which his discourse takes place or has any sense of his scale of analysis. The insights of the "linguistic turn," the shift of philosophy toward questions of language that took place in the second half of the nineteenth century and the first half of the twentieth, are here completely lost or abandoned.

So you can guess which pseudo-problems arise over the course of an ascent from the word *emerge* to *emergence* to *emergentism* to *emergence theory* to *kind of emergence theory.* The rapid, essentially unreflective and inexplicit move up the ladder of speech levels leads at every step to a new reification, which produces a series of grammatically correct but meaningless questions and answers (Wittgenstein once said, "Here language goes on holiday"[24]). What falls by the wayside are any questions about the meaning and validity of a scientific language that, like biology, seeks to explain emergent, higher system-characteristics on the basis of lower ones.

This is all a type of *discourse about science* that no longer pays attention to science's objects and plays the game of philosophy instead. It thus generates and maintains its own philosophical problems. Types of emergence are reified as objects of analysis, even though these objects do not belong to the field of scientific research, and even though they can only ever be treated as the reified varieties of different emergentisms. To say it more plainly, the elaborate appeal to science, the limitation of science to the natural sciences, is the last time the book actually deals with science at all. It is perhaps too much to hope that physicists, chemists, biologists, or other scientists could learn something about solving their problems from philosophical debates. The specific analytic style used to analyze or compare mainstream or in-group theories may call itself naturalistic, but it has little to do either with the natural sciences or with nature.

How does this study on the competition among emergence theories end? Stephan summarizes his overview of emergentist approaches in a "synopsis . . . of the logical relationships . . . between the discrete varieties of emergentism." He goes on to describe a "weak emergentism." Additional descriptive words, such as "irreducibility" or "novelty," allow him to distinguish this latter, now a "synchronic emergentism," from a "diachronic" one (71). (A system is "irreducible" when it cannot be decomposed into the characteristics of its components. A system is "novel" when it includes structures and functions that do not appear in its components or precursors. The difference between synchronic and diachronic lies in whether the differences between lower and higher system-characteristics appear simultaneously or sequentially.) The intensification of emergentist positions proceeds up to "strong diachronic emergentism," a position that, though occupying a spot in the logical elaboration of Stephan's categories, is purely fictive, because, as he notes, "it has no counterpart in contemporary debates." (This is entirely unsurprising, because such a position would require a contradiction with the entire classificatory scheme Stephan introduces, meaning that anyone who held it would have been dealt, at least in the realm of naturalistic theories of emergence, a very weak hand indeed.)

This is not the place for an entire philological analysis of this philology of emergence theories. But I do want to lay out the various types of emergent phenomena in a methodical, ordered way. My presentation will benefit throughout from Stephan's work.

Trivially, let us note that the discussion of emergent characteristics is *a discussion*. When the word *emergent* is applied to things like "characteristics" or "phenomena," these latter also need to be described in language. Emergence is thus *always about relationships between different descriptions of the objects to which it refers*. The "higher" or "more complex" system that appears via emergence must always be treated as an object of reference. Explanations of its components or of its (diachronic) development from precursor systems must be made, and then discussed.

It's crucial, therefore, to clarify (1) what object of reference is actually under discussion (even if it develops over time); (2) how the semantic adequacy and the validity of the descriptions of the reference object are grounded and justified; and (3) the relation between the two descriptions, of which we will say that one treats phenomena or characteristics that are "emergent," relatively speaking, in the other.

We're coming close to the ideas I discussed earlier. How does someone decide whether the *descriptions* of the relations between tone and melody, gear and clock, colors and paintings, calculators and calculations, or telephones and the people who use them are *appropriate* with respect to the question, *How does each of these pairs fit together?* The horizon of philosophical dogma appears again, because only logically defined relationships or causal explanations (in the sense of physical interactions) could justify this in-gathering. And once again, this philosophical tradition, so trapped in naturalism, misses the simple idea that all this is *playing out in language* and is *made by people,* and so is linked to an *order of actions.* Does not the rationally acting person, considered as a source of the problem, suggest that we ought to consider this whole issue from the perspective of *ends and means?* And doesn't the very *rationality of means and ends,* which no researcher and no research techniques can do without, also need to be thought of as part of the scientific process? *Naturalism* has less to do, finally, with scientific rationality than it does with the basic flaws of logical empiricism.

If the discourse about emergent phenomena is in fact a discourse, and if that discourse involves relationships between appropriate descriptions, then the objects that appear within it are probably not naturally produced phenomena of nature. They are, rather, the objects of scientific investigation and can only be tackled with the appropriate scientific models. The typology of emergence, that is, is a matter of *relationships among models.* And models are always models built by people. We're dealing, therefore, with the goals and capacities of model building relative to the two different levels of a system. For clarity's sake, we'll start with some technical and artificial examples. These will pave the way for a focus first on scientific questions and then on natural objects that are not produced by human beings. That accomplished, we will be able to see what it means to develop emergence theory–based, scientifically appropriate descriptions and explanations for the raw material of nature.

"Synchronic emergence" refers to the relation between describable connections occurring simultaneously inside one and the same object (at different levels of a system). The distinction between "weaker" and "stronger" emergence aims to differentiate reversible and irreversible reduction; this latter is then divided among a "logical–definitional" and a "causal" and therefore empirical type. This produces the following introductory schema:

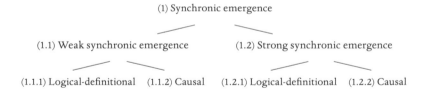

(1) Synchronic emergence

(1.1) Weak synchronic emergence   (1.2) Strong synchronic emergence

(1.1.1) Logical-definitional   (1.1.2) Causal   (1.2.1) Logical-definitional   (1.2.2) Causal

In the area of technical products, we find perfect examples to illustrate these final four types.

For (1.1.1): Let there be given two matching, interlocking toothed wheels (having a ratio of, say, 1:3). Then it is a "logical–definitional, weak, synchronic emergent characteristic" of the two wheels (aka gears) that they build a gearing mechanism. (Neither of the wheels alone is a mechanism; the higher system-characteristic of being a mechanism comes from the logical–definitional condition that the wheels interlock as gears.)

For (1.1.2): The gear mechanism is a *machine for the conversion of force.* The mechanism's power output, a transfer in the ratio of 1:3 or 3:1 according to the law of the lever (force × the lever arm = load × load arm) or the law of the conservation of energy, whereby the work in a (frictionless) mechanical machine is neither created nor destroyed, is a "causal, weak, synchronic emergent characteristic."

The two cases (1.1.1) and (1.1.2) fulfill the demand made by "weak" theories of synchronic emergence, that (1) the higher system-characteristics of either being a gearing mechanism or converting force can be worked out by returning to a *description* of the toothed wheels (this is "reduction") or can be developed from that same description (this is the "reversibility of reduction").

The move from weak to strong emergence happens via the concept of "irreducibility." We can explain it once again with reference to the toothed wheels. We will not treat them as a gearing mechanism or as a machine for the conversion of force but as elements of a very basic calculator. By placing the wheels side by side on an axle running through their centers, we can perform multiplication and division problems, so long as we restrict ourselves to problems involving a multiplier or a divisor of 3.

For (1.2.1): Our *multiplication- and division-machine* is, relative to its components, "logico-definitionallly, strongly synchronically emergent." The calculator's performance is logico-definitionally irreducible because, as far as calculators are concerned, they don't produce just any results but have to do with the production of *true or false* calculations. We're back to the simplest version of the mind–body problem: neither the geometric description of the wheels nor a series of arithmetical statements about the number of rotations they undergo can prove, even with the best of will, that "3 × 7 = 21" is *true*. The *arithmetic performance of the machine,* in the sense that it produces *valid* input–output relations, is emergent in relation to the performance of the individual wheels and cannot be logically or definitionally reduced to them.

For (1.2.2): A physical (and therefore causal) description of our little calculator would represent a system in which the movement of one of the toothed wheels, produced by some external force, transmits itself to the other one. This causal nexus is responsible for the *actually* calculated result in any given use of the machine. The *condition* for *correct* calculations is therefore the *undisturbed functioning of the machine.* Were the toothed wheels not perfectly rigid (and were thus less than ideal), if they had the properties of real bodies like Jell-O or a plastic bag full of water, then there would be a significant difference in how and how fast the larger wheel would move the smaller one, or vice versa. That would alter the calculations. It follows that despite the causal enclosure of such a system, the performance of a properly functioning mechanical calculator cannot be explained on the basis of the properties of its parts. None of the causal laws discussed here would, were the wheels malleable, be falsified thereby. But if disturbances in the system are caused by false calculations, then they do not falsify in any way the causal laws governing the movement of the pliable wheels; they just fail to fulfill their purpose. Likewise, in reverse, the causal laws do not imply (or falsify) the correctness of any given calculation. (This is the same argument, you will recall, that I made with respect to the pocket calculator.)

The shift from synchronic to diachronic theories of emergent relationships stems simply from the application of the synchronic model to processes instead of things. The system-characteristics classified as strong, irreducible, and *simultaneous* then become the *later* characteristics, which are opposed to the system components' *earlier* ones. The mysterious principle that accomplishes this transition from synchrony to diachrony comes from none other than the logical empiricist discussion of the symmetry between prediction and explanation more than a half-century ago. The "fundamental unpredictability" of qualities that were not visible in the earlier object of analysis, that thus appear or emerge later on, corresponds to the absence of either a logical–definitional or causal context between the two descriptions in the strong synchronous case.

Weak *diachronic* emergent relationships can thus be reduced to strong *synchronic* ones. In the logical–definitional case the description of the later state of affairs follows either logically or definitionally from the earlier one *and* a definition or a general proposition (for instance, one about the relationship between the relative rates of rotation of the two wheels). In the case of causal explanations the strong synchronic case brings together the later relationships and the earlier ones thanks to a causal law.

The complex discussion of different emergence theories in Stephan's book is, I would like to suggest, partly designed to produce a certain stupefied agreement, a stupefaction my deliberately simple examples aim to resist. Stephan's language expresses and borrows from the formalism of systems theory and its weirdly stilted language of entities, characteristics, systems, instantiations, determinatives, irreducibilities, and so on, for which the movement upward through at least four levels of discourse about scientific objects is partly responsible. The linguistic means whereby this takes place is the nominalization and reification of relations, relations established by none other than the descriptions emergence theories themselves have chosen to use. In short, the discrepancy between the complexity of popularized emergence theory and the actual simplicity of its content results from a lack of reflection about the nature of language in which emergence theory struts its stuff.

What's more, the plausibility according to which the unpredictability of future actions could be divided into a fundamental uncertainty and an uncertainty due to a lack of knowledge is itself an illusion. You can't resolve the problem by characterizing the lack of knowledge as either an incomplete database or a limited understanding of developmental laws. The various kinds of unpredictability cannot, in a first step, be clarified by properties of the objects about which they make predictions; you have to ask what epistemological and methodological assumptions shape the predictions themselves. Once again, we have an apparent philosophical problem produced by the use of unclear or ambiguous language, which is produced by a naive grasp of the object of description and could be resolved with some careful reflection about the scientific and epistemological means of description themselves.

This leaves us with one more ostensibly major problem to clarify. Let's consider the problem of emergence-theoretical observation within the framework of the analytic philosophy of mind. How is "downward causation"[25] possible? Example: given the assumptions of irreducibility, how can human mental capacities, which emerge from their physical ones, affect those physical capacities in turn? Isn't the principle of the causal enclosure of nature violated the minute someone throws a stone?

In a few moments I will leave behind these debates on emergence theory to take a quick look at analytic philosophy of mind and then come back to the question of naturalization of information. But first let us close this out. What do emergence theories tell us about the fundamental problem of whether information is a natural object, such that it can be characterized by a purely syntactic approach? What do they have to say about how the meaning and value of linguistic communication emerge from the technical model of a syntactic-structural system? The answer is obvious: obviously nothing. The general development of concepts of emergence in all their forms and specifications make plain the unresolved problem of methodical ordering. Consider that any given object is produced, and that discussions of that object are produced, too. This suggests—as do all of the other examples we've seen—that the object begins, not with its "basic," "elementary," or "lower" systems, components, or characteristics, but with the higher, allegedly emergent qualities of the object itself. And yet we see a certain philosophical refusal to do anything other than analyze outputs (in the form of linguistic–syntactic objects, that is, texts), rather than consider the processes that created them, the grounds and ends toward which they are directed, and, last but not least, the difference between successful uses and unsuccessful ones.

What the section on systems theory has aimed to show, and what the invention, production, and use of calculating machines of all types also teach, is the methodical primacy of the complex over the simple, the goal over the means, the effects of functions over their use, and of technical models over the description of natural ones. Emergence theory makes no real contribution here; it does not allow us to write a happy sequel to the story of the naturalization of information, in which the latter, a natural object, now develops into higher and higher communication processes, emerging in the form of a conversation on the street or of a scientific text, a poem, or (felicitous dream!) a book of philosophy. The question arises, then, whether somewhere outside this obviously *hermetically sealed tradition* of understanding the world, human beings, and the sciences anyone has sought to develop alternative points of view. And so: what do the *humanities* have to say? Hasn't the hermeneutic tradition (taken in its broadest humanistic sense) conceived of itself specifically as an alternative that would escape naturalism's false conclusions? At least it should be possible to expect something from those humanistic fields, as well as the parts of the social sciences and philosophy, that have set themselves in opposition to the naturalization of human beings, their history, and their speech. Let us go and make our visit.

PROBLEMS WITH THE HERMENEUTIC THEORY OF MIND

Our review so far of the prehistories and the forms of the myth that "information is a natural object" has carried us through the fields of mathematics, physics, and the natural and technical sciences. We also considered the philosophical side of things, looking at those schools, programs, or claims about language that have historically aligned themselves with the successes and method-ideals of the scientific disciplines. And we touched on versions of the mind–body problem, whose professions of faith in naturalism happened without any justification via axioms or other fundamental principles. That this is so means that these problems in the philosophy of mind can adopt no critical attitude toward the human-produced natural sciences, even as they draw from those sciences all their ideas and concerns (we can see this happening today in the debate on determinism in neuroscience). Where do the humanities fit into all this?

Let us not understand "the humanities" as merely the category of things left behind by the natural sciences, as you might find it used, commonly enough, by scientists themselves, but as the sciences which, in a strict sense, aim to understand texts and actions. At that point we may well expect them to generate a new information-concept for us. Whereas the natural and technical sciences and the ideals of their corresponding methods see their field of analysis as *universal,* their results as *law producing,* and their means as *the explanation of causes,* the humanities emphasize the *individuality* of their objects, the *historicity* of their results, and their means as *understanding.*

When it comes to scientific research into human beings in particular, the natural sciences focus on the species; on the type and function of an organic system; and on the natural aspects of its contexts, such as eating, reproduction, and evolution. The humanities investigate an individual, a person with a specific biography, located in a cultural–historical situation; they accentuate mental and psychological conditions like consciousness, intention, the understanding of actions, or language, all of which take place within a given cultural context. In a gross simplification one might say that the natural sciences study human beings *materialistically,* whereas the humanities study them *mentally* or *cognitively (mentalistisch).* Can we not then reasonably expect these mentally oriented humanistic sciences to provide some alternative to the naturalistic concept of information we have seen so far? Might the hermeneutic tradition, with its fundamental orientation toward understanding, grant us a starting point for the discovery of such an alternative?

Now the history of reflection on the sciences, and especially on the differentiation of the natural and humanistic disciplines, teaches us some caution when it comes to classifications like these. Two examples will demonstrate that such caution is a good idea.

In a now-famous 1894 Rectorial Address at the University of Strasbourg, the neo-Kantian philosopher Wilhelm Windelband defined the natural sciences as "nomothetic" (meaning law asserting) and the human sciences as "idiographic" (that is, as descriptors of the singular).[26] In other words, the natural sciences focus on the formulation of natural laws and the humanities on the description of individual people, situations, or events. But this division is false on both sides: every nonscientist knows that the sciences sometimes also study and describe the singular, and not just because some singularity is an individual instance of some more universal pattern or law. A look at a history of the planet Earth—think of Alfred L. Wegener's theory of continental drift (in "Die Entstehung der Kontinente und Ozeane" of 1915)[27]—would find there a scientific explanation for the origins of continents, which can be verified by looking at the form of the so-called jigsaw fit between the geology of eastern South America and western Africa. Because there is no other planet in the universe on which the exact same process has taken place, Wegener's theory is about a singular event. It is nonetheless scientific and has been proven valid. Similarly, theories about the extinction of the dinosaurs, the origin of our solar system, or the causes of the hole in the ozone layer are idiographic in Windelband's sense.

The same holds true in reverse. The humanities can be nomothetic, as, for instance, when we speak of classes or genres of literature or of patterns in political history, or when we explore regulative contexts like productivity or monetary value in national economies. These examples also remind us of the risks of oversimplified classification—as when Windelband organizes thought around the simple opposition between universality and singularity.

Another major schema for thinking the difference between the humanities and the natural sciences, operating widely in the public consciousness, comes from the English writer C. P. Snow's 1959 lecture on "two cultures."[28] One suspects that this slogan for an unbridgeable opposition between the two fields is more frequently cited than the book has been read. *The Two Cultures* offers an amiable description of different styles of living and doing research presented by an older English gentleman trained in the natural sciences rather than a solid analysis or careful discussion in any specific field (the theory of science, philosophy of language, the history of science, the sociology of science). Nevertheless, the two-cultures slogan is quite influential. Snow's distinction between the two preferred styles of thinking has in fact affected the current self-understandings of the respective fields.

One finds an impressive instance of that effect in action in two books: Dietrich Schwanitz's *Education: What Everyone Needs to Know* (1974) and Ernst Peter Fischer's *The Other Education: What Everyone Should Learn from the Sciences* (2001).[29] The first offers a canon of historical–literary knowledge, the second a history of the natural sciences. Both books rely, briefly, on the contrast between the two cultures of knowledge. Schwanitz speaks, with a glance back to the well-known Snow, of the two cultures as the "literary–humanistic culture of classical education on one side, and the technical–natural scientific culture on the other." But Snow's call has, the author notes, had "almost no effect" in Germany (482). Therefore, Schwanitz is compelled to assert, "regrettable as it may seem to many, scientific knowledge cannot be hidden, but it does not belong to the realm of being educated [Bildung]." By contrast, Fischer wants to "introduce" his readers "to the sciences" so that they can recognize and understand the ways in which they "bring meaning to the conditions of human life." Fischer goes on to argue that "many leading lights of German intellectual life . . . commonly produce only thoughtless or speechless efforts when discussing the natural sciences." He claims that this is a specifically German phenomenon. "In Germany the natural sciences and being educated [Bildung] do not necessarily belong together" (24).

Both books thus put Snow's superficial program into some kind of actual practice. But they also extend and deepen *the conflict between the natural sciences and the humanities*—even if this runs counter to the effect intended by Fischer's book, which reacts against Schwanitz's. Only at the most general level, which is to some extent given by the factual coexistence of the fields in the modern institutional university, is their possible unity asserted or simply permitted. What never appears is any kind of concrete proposal for fruitful cooperation. This contrasts with what I am attempting here, which is to build a new concept of information that, even as it overcomes the misunderstanding that thinks of information as a natural object, nonetheless can be used in the technical and natural sciences.

Having armored ourselves against a too-simple separation of the forms of human knowledge, it is now possible to ask again why the *hermeneutic* tradition, and other traditions explicitly oriented toward the *understanding of language* have not opposed themselves to the purely syntactic, materialist, and causally interpreted concept of information. This is evident in particular in the field of the newly developed media sciences, and in their philosophy. There one finds, with very few exceptions, an understanding of information and communication, of nature and of technology, that essentially follows the one laid down by the natural sciences. And the tradition of the "analytic philosophy of mind," which led a broad debate about specifically human structures (by means of themes like self-consciousness, intentionality, and qualia), has never launched itself into reflection on the tacit premises of the approaches to language entailed by either the semiotic tradition that follows Charles Morris or the idea of natural science as the Vienna Circle understands it.

A quick look at the history of hermeneutics shows that, as an art for the interpretation of texts, it has always oriented itself toward *clearly assignable goals*. In ancient times the art of poetry (poetics) as a technique was distinguished from knowledge, politics, and ethics, and worked only in their service; so too the poet's interpretation served the moral and political perfection of the person. With the advent of scriptural Christianity, hermeneutics acquired a new task. While the books of the Old Testament had been directed only to the Jewish people, the New Testament aimed to reach all of humanity; it required, therefore, new reading practices that would give a reasonable coherence and consistency to the holy texts. Following Origen, Saint Augustine proposed three types of interpretation: a literal one, oriented toward historical and grammatical analysis; a psychological one, oriented toward morally relevant interpretation; and a pneumatic or spiritual one, which would describe the holy text's allegorical and mystical dimensions. A millennium later, Martin Luther made it a principle of scriptural interpretation that the biblical text be read as the verbatim word of God. From the beginning of modern times, a new problem arose: the demand that philosophical (including scientific) texts be understood according to the principles of reason. As biblical texts, under the influence of writers like Friedrich Schleiermacher, began to be interpreted as historical documents (as products, that is, of human authors), the hermeneutics of the twentieth century developed an exegetical method that was influenced by a number of different background philosophies. As a result of these changes, the interpretation of texts and the understanding of discourse lost sight of their earlier interest in rational action and definable goals.

Despite all these historical developments, one influence from ancient philosophy remains: what one might call the *banausos verdict*. In *The Republic* Plato had granted craftspeople (Greek: *demiurgos, banausos*) the same status as slaves and women, separating them from the *polites*, the free citizens of the state. The typical *banausos*, as explained in Aristotle's *Nicomachean Ethics*, lives under the goal-oriented rationality of production *(poiesis)*. When a carpenter makes a table or a bed, he does so neither for his own pleasure nor as an end in itself but because of the utility of the bed or table. And poetry was likewise, as I mentioned earlier, no "theory." It was not knowledge earned by looking at things from a distance or by assiduous reflection; it was a technique, an ability.

It was particularly German-language philosophy that bound together this contempt for goal-oriented rationality and technique (*Banausenverbot,* an interdiction on craftspeople) with a demand for psychological and emotional understanding and empathy. In doing so, it denied to hermeneutics an entire method that was, by contrast, at least stylistically obvious to natural scientists trained to attend to experimental experience. While the operationalization *(Operationalisierung)* of basic concepts is de facto unavoidable in the natural sciences (and other philosophies, especially of formalist origin, have been developed to consider it), for the traditional German-speaking humanist, such an operationalization remains a deeply foreign idea.[30]

The telecommunications approach to the concept of information, and its orientation toward the carrying capacity of a given channel, has been, as far as the structural preservation of transmitted signals goes, a supremely successful operationalization. One can also easily recognize operationalization in the applications of the information-concept in the natural and technical sciences. But the classically humanistic aspects of a person—consciousness, intentionality, the understanding of language, the ego-perspective—have not been the objects of any definitional efforts from the humanities themselves, even though they are a kind of operationalization, minimally insofar as they correspond to a reference to the activities of daily life.

There thus exists—I concede the generality of this judgment—a certain parallel between the sciences and the humanities with regard to the care they have for their *own* basic concepts. The materialist defines what matter is about as stringently as the mentalist defines the "psychological." The empiricist, deriving from his experiments causal explanations, and formulating natural laws, defines or justifies the causal principle, the experimental method, and the legal structure of natural laws about as stringently as the hermeneutic scholar explicitly defines basic methodological concepts like "understanding" or "empathizing," "self-consciousness," "intentionality," or "ego-perspective." In both cases everyone assumes that these specialists already know what they are talking about.

This style can be found also in the analytic philosophy of mind, even though it stems from other traditions. There too principles are taken for granted and basic understandings as given, as you see when you compare the analytic methods used in emergence theory, about which I wrote earlier. Despite recent efforts to naturalize self-consciousness, intentionality, and the ego-perspective, there too you will find no attempt to operationalize the concepts at hand. There is no real reason to say that the analytic philosophy of mind has been influenced by the ancient Greek ban on *banausoi*. At most one could say that it is as if goal-seeking rationality and operationalization have been, for whatever reason, prohibited there, that the traditions from which analytic philosophy grows, because they naturalize human action, altogether *ignore action* as a source of knowledge. Analytic philosophy lacks, therefore, a pragmatic and epistemological turn. We can therefore speak of a failure of hermeneutics, owing to its lack of interest in operationalization, that applies both to analytic philosophy and to the traditional humanities (especially the German-speaking ones) more generally. The mentalists have little reason to criticize the materialists, either on language-philosophical or epistemological grounds.

FIVE
METHODICAL REPAIR WORK

To repeat: my critique does not deal with the scientific fields involved in the naturalization of information; it addresses their accompanying philosophies. There is no specific scientific question, What is information? that could be answered at the level of the language of scientific content or objects of scientific research. The question poses itself, in certain circumstances, at the metalinguistic level at which scientific claims and results are talked *about*. On closer inspection, however, the assessment of the information-concept also includes, besides those statements by scientists operating at the level of an object language, the various modes of metalinguistic determination, in which for example one sentence operates as a definition, another as a principle, and a third as an empirical result. So answering the question, What is information? requires, in fact, thinking of a third level of language, in which the very distinction between what counts as objective language about information and what counts as metalanguage about it takes place. Unfortunately, debates on the question so far have not been carried out with the requisite care that would allow the various statements to be considered with respect to speech levels and the criteria of value associated with them, or even to be recognized as belonging to such levels at all. The "philosophizing" in much philosophy that goes on around science amounts, unfortunately, to an undisciplined, monologic babble.

Where the concept of information is a matter of debate—when two representatives of different points of view battle it out for their respective philosophical positions—the self-images of science and scientists all too often are represented as the quasi-inevitable *consequences* of the respectable results produced by scientific disciplines. In this way, each side will claim the scientific status of disciplinarily serious results, declaring itself free of the kinds of emotional coloring or bias inherent to any self-understanding. But it can easily be demonstrated (and in fact has been demonstrated, in the preceding chapters) that the philosophies that accompany the naturalization of information cannot be derived from the results of scientific inquiry. It makes more sense to say, therefore, that the self-understandings of certain popular philosophical positions stem from decisions about fundamental principles that are not themselves consequences of scientific research but rather matters in which that research is *already invested*.

Any attempt at a general repair of the methodical situation of information theory will, of course, have its own investments. I am not interested in producing an entertaining fiction, with starring roles to be played by arguments that make, miraculously, no assumptions, and ideas that, incredibly, produce their own foundations. Accordingly, this effort to fix things will (1) *begin with the common practices of everyday speech and ordinary life,* which for good reasons cannot and should not be neglected. These practices establish a starting point for thinking about which goals the search for a reasonable discourse *(Rede)* on information might establish for itself. The first decision to make is to differentiate *human communication from natural processes.* And since we are, rightly, made responsible for our own discourse, I will (2) *treat communications (Reden) as actions.* Thinking this way brings to light, particularly with respect to action sequences and the actions that respond to them, the prediscursive assumptions that motivate every communicative event.

Being responsible for one's own communication means, also, *talking about communication in a methodical,* nonchaotic way. Only after these philosophical reflections on the investments that underlie my argument, and a detailed elaboration of the basic program, is it possible to (3) explicitly lay out the discourse on information. That means, first, defining the most important foundational concepts of information theory as well as their language-philosophical components. Once we have an appropriate terminology in place, it will be possible to discuss the two most important aspects of the career of information, namely, (4) the technical systems that aim to substitute for human communication (beginning with the writing down of speech) and (5) the history of successful techniques for the modeling of natural processes.

Customs and Purposes as Investments

Beyond their use in science and philosophy, the various words and distinctions subsumed under the subject headings of communication theory, information theory, and communication technology also do *indispensable service in everyday language.* Someone who looks up the train schedule is hoping for reliable, understandable information. He seeks knowledge, a set of facts, that will be useful for future action.

The familiar and natural engagement with the seeking and finding, the giving and taking of information, includes the expectation that the daily business of reciprocal informing will involve misunderstandings, errors, deceptions, gaps in knowledge, lacks of clarity, and the like. In other words, *the familiarity with the untroubled normal case* of reciprocal informing does not rely on explicitly defining the normal case as such. One only becomes aware of the normal case as normal when it is in some way disturbed. *Disturbances* occur vis-à-vis the unreflectively performed normal case. Because such disturbances are a nuisance and should be avoided, a *science* of reciprocal informing develops with the aim of managing its errors. Sciences designed to support the practices of the living are, therefore, *disturbance-avoiding, disturbance-removing* forms of knowledge. To accomplish those tasks, they must define for themselves an *undisturbed ideal situation.*

A methodical approach begins, then, with parts of a living practice, which take place and are lived in an undisturbed normal case. Science then defines this case as the *ideal, undisturbed case.* In any lived process, disturbances produce a knowledge appropriate to their management, a know-how, that allows us to avoid or rectify them in the future. To describe that know-how will require finding at least *one kind of language* that will not itself represent a disturbed case and that might therefore prove, given sufficient explanation and definition, a reliable means for the production of knowledge.

KNOWLEDGE TRANSFER

If I have been using the stylistically inelegant formulation "reciprocal informing," it has been only in a provisional way, because of course not all communication (meaning, a discussion with another person) is informing, in the sense of being a transfer of knowledge. Sometimes people invite each other to things; they ask questions or perform narrowly defined speech acts (petitioning, blaming, praising, promising, naming, etc.); sometimes they speak to themselves or just say aloud what they feel. We may well ask of all these so-called expressive speech acts whether their primary goal is to inform another of something or whether they are not manifestations of other needs—concealed appeals for help, calls for compassion or wonder or anything else—that might best be characterized as solicitations rather than as descriptions or claims (with the proviso that such avowals can be genuine *[wahrhaftig]* but not true *[wahr]*). So the formulation "reciprocal informing" reminds us that we are discussing only one aspect of linguistic communication, the aspect that deals with facts, knowledge, and their value, and which therefore belongs to a general field involving the judgment of truth or falsehood.

I am now going to leave aside the question of why it is appropriate to acknowledge the common operations of ordinary language, with which we inform ourselves and say that we inform ourselves, and about which we know that they always run the danger of producing misunderstandings and other disturbances. I have no ambition of getting us to the point where we can imagine a daily life that would not include the possibility of reciprocally informing one another. Almost nothing that we know comes directly from our own learning; we acquire almost everything from others. Yes, it is possible to say that, at best, only in the realms of personal, individual, or even intimate life does what we know come from our own experience (apart, let's say, from the production of knowledge in successful research). But as soon as knowledge becomes public, as soon as it subjects itself to validity claims, it belongs less to our own personal cognition and more to a general borrowing from third parties.

Something does not have to be scientifically known to be universally known. We expect telephone directories and bus schedules to be correct; otherwise, they're worthless. The same goes for atlases, geographical and political information, knowledge about nature and history, and really any kind of general knowledge (including the sciences themselves). The entire field of human discourse that primarily concerns understandable and valid *claims* is indispensable to the management of life in all its forms, from the private to the public to the transsubjective–scientific and the technical. And this discourse is first and foremost a process of informing, of giving and taking of knowledge, facts, and ideas *(Wissen und Kenntnis)*. That's why any attempt to repair the naturalistic misunderstanding of the information-concept must begin with the common practices of everyday language.

THE ENDS OF COMMUNICATION

Traditional theories of information and communication do not ask themselves why-questions. Informing and communicating are always already taking place, and as natural processes besides. From a cultural point of view, however, we humans are naturally also always already discursive beings. We do not only begin communicating because some desirable goal compels us into speech.

It is therefore helpful, if we want to clarify the concept of information, to ask why-questions about practices that have long been taking place. In this way we can retroactively verify whether communicating and informing constitute adequate means for the ends they aim to fulfill.

In what follows I want to understand *communication as a tool,* as a means *of organizing cooperation.* This may sound somewhat technocratic, so it's worth saying that I am speaking here of morally serious communication that shapes the situation of living beings and not about the style of chatter or loquaciousness that belongs to the world of undemanding entertainment or idle amusement.

Human life relies on cooperation. This is true not only in advanced industrial, complex societies but also in all those familiar, minor forms of community, which may allow us earnestly to speak of a cooperative being-together. The basic dependence of human beings on one another, in both the culturally significant learning phase of childhood and the inevitable struggles of old age, offers us a good reason to insist that the goal of communication is rational discourse, if we understand the latter to refer to fruitful cooperation. It is precisely this dependence on cooperation that provides the justification for *making ourselves reciprocally responsible for our discourse.*

The Responsibility for Discourse

We become responsible for our discourse. This happens naturally when it comes to relations between people. We "hold someone to her word" in any number of ways: when we promise or plan something, when we find or end a friendship, when we found or disband an organization, when we sign a contract or place an order. And when words like "please" and "thank you" are not communicated gesturally, we express the responsibility for discourse directly and literally to the person whom we have asked for something, or thanked. *Language assigns either merit or blame to those who use it.*

In so doing, communicative discourse fulfills one of the central definitional characteristics of action. *Communicating is acting.* Actions can work or not work, be successful or fail, depending on whether they reach or fall short of achieving their goals; they can be abandoned, they can be responses to a request, and many things more besides. We are made responsible for our discourse because discourse is a special form of action, and action differs, in turn, from a natural (that is, innate) behavioral repertoire insofar as we have to learn it, laboriously, within the context of a speaking and acting community.

Seeing things this way offers us a first set of rigorous objections to naturalizing appropriation. Informing and communicating should not be connected from the beginning with a wide array of metaphors and associations (like the idea of channels of communication). When the sun warms the stone, no one needs to say that the former is the sender and the latter a receiver, or that the sun informs or communicates something to the stone. It's equally weird to say that the warmed stone is a sign for sunshine or good weather. Theoretical discourse about information and communication ought *programmatically to limit itself to human language,* because only for language—where the concepts are in fact primary—does it make sense to talk of meaning and value.

Nothing about those claims forecloses the later use of language-philosophical clarification as metaphor (for instance, in the description of technical or natural processes). What is determined here, however, is that the distinction between *made* and *found,* between *technical* (or *artificial*) and *natural,* between *culture* and *nature,* emerges from the world of everyday life as a primary distinction. Someone who finds a mushroom in a forest assumes it grew there naturally; that same person, finding a car key, takes it for a lost artifact of human culture. No theory of the world or of humanity can be foundational if it is the consequence of an already-established, human-made, and historically developed science. Materialist and naturalistic philosophies are, in this view, ungroundable and illegitimate.

ORDERING ACTION AND DISCOURSE

These programmatic statements are neither arbitrary nor dogmatic; they invoke the needs and goals of a thriving human coexistence. Only by making ourselves mutually responsible for our discourse, and by understanding communication as action, do we make it possible to understand historically developed technologies on the basis of their purposes and to grasp their role in the modeling of nature. And only by making ourselves mutually responsible for our discourse do we allow the *principle of methodical ordering,* which my earlier attempt at *methodical* repair aimed to bring into being, to play its proper part. This principle forbids us to speak of a sequence of actions as though the order in which they take place were not a component of their success, at least when we are talking about the consequences, nature, or social function of that success. In everyday life, everyone understands the idea. Think of directions: no one appreciates a user's manual, set of assembly instructions, or recipe that, because it lists the steps out of order, leads to failure. Similarly, no one enjoys newspaper reports or narratives that tell things in a completely disconnected way, so that the events are implausible or make no sense. The principle of methodical ordering shapes the way we talk about action, descriptively as well as prescriptively. That acting often involves chains of actions, so that the purpose and success of each individual action depends on its being part of a chain, and that those individual actions cannot usually (but not always!) be exchanged without affecting the success of the chain as a whole gives the "rational" sequencing of discourse about actions a *pragmatic criterion of judgment.*

If communicating is an action, it becomes possible to wonder in which cases the principle of methodical ordering applies to discourse itself. Discourse about discourse must also follow the principle of order, at least when discourse represented as a discourse is bound to a sequence that allows it to fulfill its purpose.

The action-oriented character of discourse gives the latter certain traits that theories of communication and information have regularly overlooked. They allow us to distinguish between two aspects of human action.

FUNCTIONING AND SUCCESS, COLLECTIVELY

Action can be thought of as *participatory, communal,* or *individual.* Participatory action is action that can only succeed thanks to the participation of another actor. A classic nonlinguistic example would be a race or a tug-of-war. To take place, both cases require the freely chosen participation of some other individual. One might say, being a bit careless about the intentionality of speech, that participatory action takes place only when it is possible that its participants can establish an agreement that articulates a prediscursive, preactive consensus about the action to be taken.

In this way we can distinguish participatory action from communal action, in which the help of another person is critical not so much to the taking place of the action as to its overall functioning or success. A beam too heavy for one person to lift requires at least two people, each of whom knows for herself what she has to do. The action only succeeds when two people work independently together.

Other actions, neither participatory nor communal, belong to the class of individual actions, which a single person alone can put into practice and make function or succeed.

Without getting into questions of empirical evolutionary development, one can nonetheless observe that the acquisition of the capacity for action or discourse for a human individual can only take place *in communities of discourse and action* in which *participatory and communal activity plays a fundamental role.* The raising of children, all the collective processes of teaching, demonstrating, imitation, praising, and correcting, indeed, the entire social process that attends to newborns, infants, and children, is enough to demonstrate this truth at work. And this applies more generally for the actions of communicating and exchanging, so long as we do not limit the latter to the artificial situations of practicing a monologue or a proclamation. Communication is a form of participatory action, because it can only take place when at least two communicating beings follow an agreed-upon protocol for doing so (or, alternatively, work to create one—imagine two people who don't speak the same language using gestures to establish the ground for future exchanges). But for communication to produce results, that is (in an undifferentiated sense), to succeed, it must also be understood as a communal practice.

OBSERVERS, PARTICIPATIONS, AND IMPLEMENTATIONS

A second aspect of action leads from action theory to an important distinction regarding discourse. It lies in the *differentiation between implementation and description,* with this latter capable of being split once again into the well-known division between *participant* and *observer perspectives.*

When the distinction between participant and observer perspectives was introduced to debates in the social sciences, the philosophy of sciences, and ethics, it was seen, quite rightly, as a significant step forward in our understanding of knowledge. I will skip an overview of its history and simply say that it was an important insight that efforts to do research in the fields of historiography and social theory or in theories of norms suffered from the fact that the authors of the work were at least partially implicated in the ideas they developed, that their perspectives were shaped by the fact that they were *participants* in their own research. Only where an author could remain completely outside the context of research would it be possible to say that he had acquired the perspective of an *observer.*

But here we arrive at the epistemological punch line: the adoption of the perspective of a participant or an observer is not a matter of arbitrary or free choice; it is, rather, the necessary condition for the success of any endeavor to know or justify anything at all. An example: it makes no difference to the truthfulness of an anatomical description of the human skeleton that the anatomical scientist herself has a skeleton. From that point of view, anatomy can view things from an observer's perspective. The field of sensory physiology, which does research into the act of seeing (the construction and function of the eyes and the visual cortex in the brain), has on the other hand an object, and a method for studying it, that can only be spoken about from the perspective of a participant. After all, someone who does not know what seeing (in the ordinary sense) "is," someone who cannot come to a consensus with other sighted people about what can be seen in the results of an experiment in sensory physiology, will not be able to join the ranks of sensory physiologists. The very objects and methods of a highly prestigious *(angesehenen)* natural science (!) force its researcher to adopt the participant's perspective, even if, in the history of the naturalization of the sciences, such a fact has been completely covered over. Whether you speak of the eyes of an octopus, an insect, or a human being, you will end up using a metaphorical translation of the idea of "seeing" as humans experience it.

It's significant that both perspectives, those of the observer and the participant, are *perspectives from which description takes place.* Applied to speech acts, we would then be speaking about qualities of speech that can really only be handled by metadiscourse (either as a participant or an observer). But the participant's perspective does not open up those aspects of discourse that are reserved for the perspective of the *implementation (Vollzugsperspektive)* of the speech act.

Unless we are talking about the action of describing, in which case description constitutes itself as action, description alone does not bring action into the world. Actions have to be *implemented (vollziehen)* to be actualized. Implementations *(Vollzüge)* thus play first and foremost a significant role in bringing discourse (and therefore language) into the world through discourse, and not through theories or descriptions of it. Likewise, the objects of the theories of information and communication do not come into being through the efforts made by the folks who theorize them. Already in the history of ideas that leads to the naturalization of information as it did in the semiotics of Charles Morris, the idea of implementation *(Vollzugsperpektive)* is completely hidden, well-nigh inaccessible. Only from such a position could the examples of a dog chasing a squirrel and a person receiving a postcard be thought of as identical. The dog, by definition, *does not act*; it does not carry out an action of any kind. Whether the recipient of the postcard does so depends on the situation. It could be a matter of pure happenstance, something that just "takes place" for him.

The distinction between participant and observer perspectives can only be made by reflecting on the way that each perspective alters the validity of the knowledge that derives from it. We must think the same way about the implementation perspective. The double meaning of implemented actions lies in (1) the *irreducibility of the implementation to its description* and (2) the role of the *implementation as a starting point for the justification of knowledge.*

Faced with a descriptive withdrawal of the *author* of knowledge, whose programmatic spiriting away and naturalization constitute the very ground of naturalist epistemology, one must take into account the fact that knowledge and science are *produced* by actions. Only implementations produce knowledge. In justifications of knowledge (or refutations or errors), therefore, implementations will assume the task of *acting as starting points for justification (or refutation).* Clearly, in the case of doubts or controversy, the actual actions of the authors make available to all the facts or objects under dispute. There are examples of entire scientific and epistemological chains of such justifications. When geometry, for instance, is operationally justified, you find yourself no longer *talking* about the description of the actions of producing spatial forms or tracing figures; rather, these actions are *actually concretely carried out,* operationalized, because otherwise the objects in question would be inaccessible and their hypothetical fiction would lose any relation to reality. Implementations make the objects belonging to the reality in question available for use.

DISCOURSE AS ACTION

What goes for the nonlinguistic, poietic actions in a context like that of Euclidean geometry goes equally well, language-philosophically, for discourse. The primary object of theories of communication and information, on whose grounds those fields have to build themselves up and against which they are tested, are implementations of speech acts. It is on these grounds that the criticisms I made earlier regarding the metaphorization of information and communication in the case of molecules and neurons acquire their full sense and force. There are no implementations there. Their fictionalization via metaphorical description has no objective counterpart. And this thesis resolves itself from the implementation perspective. Biology, which speaks of genetic information, does so to make the language of molecules the basis for understanding the language of human beings. It puts itself into a *performative contradiction* with regard to its own language implementations in the very production of this theory.

At this point it may be clearer why it matters that humans make other humans responsible for their discourse. We become *responsible for our implementations* because they have consequences that would not be in the world without us. The consequences of speech acts, whether they themselves take the form of speech acts or not, are just as much a matter of responsible action as are the consequences of nonlinguistic actions involving movement or making. Just as we hold someone responsible for a punch (a movement) or for painting, or setting fire to, an oil portrait (forms of making), so we hold him responsible for his speech acts, in everyday life and, a fortiori, in philosophy and science. When theories of communication and information are sustained by false philosophies, they end up as forms of thoughtless and distanced descriptivism that are scientific in name only. The distance of the scientist from her object cannot be eliminated from the science of discourse, since the implementation of discourse itself falls under the umbrella of the science that studies it. And in any case, implementations of discourse *(Vollzüge des Redens)* enter the world not through the science of discourse but by being implemented. The specific achievements of the sciences must emerge from those actions if they are to acquire transsubjective value.

Now that the language-philosophical fundamentals of the various theories of information and communication have been gathered together, it is time to attempt a few terminological clarifications.

Communicating and Informing, Terminologically

A *methodical opening* toward the construction (or reconstruction) of a relevant terminology has to begin with communication, not information. No person has ever learned to use language under the stylized conditions imagined in Shannon and Weaver's mathematical theory of communication. As I suggested in my critique of that theory, communication takes place only by virtue of a *continuous exchange of roles,* because only in that reversal can those who communicate evaluate whether their speech acts are either functional or successful. Actual discourse between human beings features no external, Archimedean observer capable of describing and deciding whether the communicators have understood one another, have mutually recognized each other, and so on.

UNDERSTANDING AND ACKNOWLEDGMENT

The understanding and acknowledgment of the speech acts directed by one person to another can be thought of in terms of the differences, taken from action theory, between functional and nonfunctional, or successful and unsuccessful, discourse. (We thereby set aside analytic language philosophy's competing theories of meaning and truth. I develop a methodical philosophy of language, derived from the classic distinctions of a methodical theory of action, in *Logisch-pragmatische Propädeutik*).

Beginning again from the common practices of everyday shared discourse, I assume that every speaker wants (1) to understand the other and (2) to be acknowledged as the producer of a speech act. To leave behind, as a first gesture, the entire field of philosophical theories of truth, consider a simple request. What does it mean for *a request to be successful* (in the sense derived from an action-theoretical distinction between functioning and nonfunctioning, in which the speaker primarily judges whether her action is "correct" in the sense that it expresses her intention, or whether, in the cases of regulated practices from figure skating to chess, it has followed the rules of its practice)? A nonfunctional request would take place when, as the result of a slip of the tongue or a mistaken gesture, the request becomes incomprehensible for its addressee.

Just as, in the case of a nonlinguistic action (like a movement or the making of something), to "understand" an action means that the participant recognizes *which action-schema the actor implements (vollzieht)*, a request is successful when its addressee feels that something has been requested of him. (For purposes of transparency we are dealing here with the undisturbed normal case, in which the request's addressee also recognizes *which action* has been requested of him. The easily imaginable situation in which a person feels that something has been requested, but does not know what exactly, is passed over here for simplicity's sake.)

Of course, it "takes place" that the speaker wonders if the addressee has understood his request. That means that the functioning or nonfunctioning of a given request has, for the requester, an action-oriented as well as an *eventful character*. The requester cannot arbitrarily enforce the understanding or misunderstanding of an invitation; understanding and misunderstanding are effects of participatory communication. In the simplest case, in which the requested person wordlessly obeys the request, the speaker observes that she has been "understood." What it means to understand or to misunderstand here leads us back first to the functioning or nonfunctioning of invitations in particular and can then be applied, mutatis mutandis, to every other type of speech act.

In learning a language, we also learn the schemas whereby one usually "reacts" to invitations, questions, or claims and whereby one reacts to the speech acts of greeting, asking for something or thanking someone, insulting or promising something to someone, or expressing mental states. At the same time, as speakers, we learn how to understand the various cultural practices governing these reactions themselves. Someone who gets the answer "five o'clock" to the question "how do I get to the train station?" has probably, most of us would feel, been misunderstood.

One judges the success or failure of a request when it is either followed or not. In the former case, we will speak of the *acknowledgment* of the request. In general, all types of speech acts entail particular forms of acknowledgment or nonacknowledgment. The acceptance of a claim constitutes the most basic step of acknowledgment on the road to a discursively derived truth.

To recapitulate: someone makes a request. Its understanding produces a response, an adherence to its terms; this is acknowledgment. The success of the request, produced through these processes, constitutes its *validity*. Similarly, then, we may define the failure of a request as the rejection of its consequences or instructions. The philosophical point is that the success of a request takes place for the speaker inside the context of cooperation and communication. It befalls him there, and not anywhere else.

This brief sketch (about which you can read more in *Logisch-pragmatische Propädeutik*) aims, by reviewing the meaning and validity of linguistic requests involved in all types of speech acts, to develop language-philosophical distinctions, specifically the functioning and nonfunctioning, success and failure, that characterize speech acts in cooperative and communicative situations. In other words, the meaning and validity of a linguistic expression implemented via speech acts derives only under the condition, and therefore methodically primarily under the condition, of two people who implement participatory and communal actions of communication. Crucially, the partners in a conversation can only succeed and be functional via a constant exchange of the roles of speaker and listener.

Where such practices of communication and cooperation are fully established, it becomes possible to talk about *informing*. "A informs B that *f*" describes part of a total context of communication in which a speaker gives a listener the knowledge of some fact *f*. Facts are represented in statements. The statement form is, however, not limited to the special case, for which it's merely a matter of sharing true statements and therefore existing facts, as in the case of the (correct) bus schedule. With a little artifice and effort, all types of speech acts can be reproduced in this "A informs B that *f*" structure—invitations, greetings, questions, what have you. All speech acts can be iterated as instances of "informing." "A says to B, 'Come here!'" becomes, in this context, "A informs B that she should come here."

FROM INFORMING TO INFORMATION

In the regulated contexts of ordinary public life we will encounter any number of examples in which we expect and demand from informing a *certain invariability*. To stick with the train schedule, it seems fair to say that *what* information is given depends neither on the traveler nor on the person in the information booth. Whether the speaker discourses beautifully or haltingly, whether the hearer gets things right the first time or needs repetition—these things should play no role in the giving or receiving of information. Information should be *speaker-, listener-, and representation-independent*; the words or sentences that convey it should play no decisive role—to speak a bit naively—in the production of their "content." What is the "content" of a fact? The "content" in a wine glass remains the same if you pour it into a beer mug. The transmitted discourse of the content of an act of communication should thus be *independent* from the form of its vessel, that is, independent from the specific language that transmits it and from the participants in the conversation that includes it.

In the history of culture, the demand for invariability despite variations in speakers, listeners, or representational forms of communicative acts comes from the field of the law. The fairness of laws and their application in the machine of justice consists essentially in the idea that everyone faces the law as an equal, that it makes no difference who one's judge is (or who one is) to the ways that legal norms are interpreted or put into practice.

To borrow language common among philosophers of language and logicians, we might then say that the concept of information emerges from the logical process of abstraction as it applies to the action predicator *informing*. I do not want to get too deeply into that idea here, but the following examples might serve as illustrations of its plausibility: just as different signs can represent the same number, different words the same concept, and different statements the same fact (and different true facts the same state of affairs), so can *different linguistic utterances represent the same information*. They are *informationally identical,* just as a Roman numeral III and Arabic numeral 3 are "numerically identical" or *necktie* and *cravat* refer to the same thing. In this way we can bring the *concept of information* back to the *action of informing,* which, as we have seen, can be understood as the result of speech acts made within processes of communication.

THE MECHANIZATION OF INFORMATION

Our look back at the incredible historical success of the technologization and mechanization of discourse now arrives at an important second step: relations between humans and machines. Our highly technologized modern life confronts us with any number of instances in which, metaphorically speaking at least, *humans make requests of machines,* or in which *machines make requests of humans.* In the first case, you might consider the way a car responds to its driver; in the second, think of the way the driver herself responds to traffic lights. These relations apply not only to machines that produce semantic or symbolic activity (like calculators or computers) but effectively to *all technical apparatuses that control or regulate action.* We might well talk about, for instance, the steering of a paddle boat or sailboat on its course (relative to a human destination, chosen to meet a certain goal) *as if* the boat were responding to requests made by the person at its helm (or, in the case of a failure, as if it were refusing to respond). In the reverse case of a request by a machine to a human, the states of the boat's system, like its deviations from the helmsperson's desired course, can

METHODICAL REPAIR WORK   143

be understood as requests for course corrections. The same thing for stoplights, whether these are controlled by a person, a timer, or a traffic computer. The *metaphorical* handling of human–machine relations happens as a result of a certain pattern of ascription (not description!) that acts *as though artifacts were the originators or addresses of requests*. Strictly speaking, such metaphors are superfluous. They amount to little more than a description in which the user of the machine treats and portrays the latter as a means for goals, whether his or another's.

Requests made between humans and machines begin to matter for the concept of information (and are therefore not superfluous) when control and regulation processes depend on communicative and cooperative practices, when they "depend on information" contained in the human's request to the machine or the machine's request to the human. In this situation, the *evaluation* of functional or nonfunctional, successful or failing, requests between human and machine actually works as an analogy of what happens in communication and cooperation among two people.

An example, for clarity: consider a voice-activated dictation machine, one that turns on automatically when someone speaks into its microphone. Obviously the machine will turn on even if the person says into the microphone, "Don't turn yourself on!" Unlike machines equipped with speech "recognition," which "respond" to various spoken commands, the dictation machine does not "understand" the request. This gives you some sense of whether an information-theoretical description of a machine process might be useful or whether it simply produces a redundant metaphorics.

The functions of machines can again be differentiated on the basis of instances of failure. The functional criteria of the dictation machine that "does not understand speech" do not require it to make any distinction among what it "hears." Whereas for the machine that "does understand speech," it is the failure to make a distinction (between "turn on" and "don't turn on") that defines a malfunction. The malfunctions of the speech-"recognizing" machine can only be judged and explained by a human being competent to make judgments of semantics and validity.

An information-theoretical mode of description for human–machine relations in control and regulation processes is, therefore, redundant if it does not functionally follow the understanding–acknowledgment model developed here. But even in nonredundant cases where machines are built, optimized, and used as performatively equal replacements of human capacities, the work they do is determined by understanding and acknowledgment in a context of cooperation and communication. The methodical primacy of the linguistic competence of a human speaker holds even for technologically successful speech "recognition" systems. The discourse of machine performance is metaphorical.

EFFICIENCY AND INFORMATION COSTS

Even when it comes to technology-free human discourse, the efficiency of forms of communication can be thought of in terms of cooperative processes. Increasing efficiency in complex actions—sailing a large ship, organizing a hunt, fighting a forest fire, military operations—is conventionally a matter of establishing *standard commands* for certain specific actions. A given *repertoire of action-schemas* defines the representational invariance of the commands and thereby determines the informational identity of the language (the communication acts) that express them. Any number of prescientific, ordinary and everyday examples demonstrate the need to assess the costs of communication processes; in any of these examples the given repertoire of action-schemas is supposed to be effortlessly accessed through a system as simple and resistant to error as possible. In such a situation, even communication that proceeds without any technical assistance allows for a quantitative comparison of the error rate in the invariance of information-content to be defined and established.

If we move on to processes of technological substitution, and limit ourselves to nonredundant, informational cases, it now becomes possible to reconstruct how one might speak of an "information source which is producing a message by successively selecting discrete symbols," as Shannon and Weaver do (12). The idea stems, somewhat naively, from the fact that sign systems like the alphabet or the list of words in a language present us with finite, preestablished sets of signs. This stipulation is then confused with the specific limitations of coding and decoding machines (like the capacity of a telephone or microphone to register differences in sound pressure or frequency) to, for example, their physical parameters.

To recognize this means that at this point it becomes possible—

through the methodical construction of a theory of communicative and cooperative conditions as kinds of human action and the clarification of what processes we can think of as performatively identical, technological replacements of human processes of communication and cooperation—to say *explicitly* how the fundamental and heretofore undefined concepts of information theory may be *defined* and how the normalization of conventional actions and standardized commands (the "sign system") is *legitimized* by the purpose of the cooperation (e.g., firefighting) that calls it into being.

We can now connect the definitions of information-measure, channel capacity, and so on, that follow from the Shannon–Weaver theory to the program we aim to accomplish here. So for instance, one can take the common term "information processing" and put it back into the context of the reversal of roles in speaker and listener. When person A asks person B how to get to the train station and person B responds with directions, person B has changed places from listener to speaker. The actions B takes (in listening and responding) correspond to those necessary for an identically performing machine (e.g., an automated ticketing station). We may think of the input and output (via speech or keyboard) here as *data* (givens), whether we are talking about a person asking a question or whether a machine (metaphorically) "answers" one. *Data processing* is the technical or machine-theoretical aspect of *information processing*. But the processing of information remains firmly bound to human actions and capacities, even when data processing technologically substitutes for it.

Finally, we can address the naturalistic discourse of *signs* in Morris's semiotics (and indeed in all of information theory) by returning to their origin in human communication and cooperation. People can *give each other a sign*. This requires that there exist a sufficiently familiar social convention that establishes what certain *actions of showing mean* (that is, a *schema* for making and interpreting such actions). In the trivial case of a fork in the road (*trivial*, from the Latin *trivium*, "three-way"), the person who asks "which way to X?" expects a gesture pointing either right or left in response. This kind of communicatively and cooperatively *successful action of showing* becomes conventionalized; in this sense it belongs to an action-schema. Such actions can be substituted technologically, for instance, by a signpost. Because the signpost "functions" like a person who gives directions, that is, it fulfills the same goal for the traveler, one also refers to the signpost as a "sign." This is not a matter of logical consequence.

Equal Performance and Technical Imitation

Up to this point, it's been fairly clear, despite the lack of an explanation or definition for the idea of "equal performance," how our talk about human–machine interactions translates familiar ways of talking about human actions and applies them to machines. The driver who operates his vehicle and follows the direction given by a stoplight acts "as if" he is making requests of a machine, or as if a machine is making requests of him. Is this as-if mode of speech appropriate in cases where a machine has as its purpose to function *as if it had mental capacities, like the ability to understand or produce speech*? How can we determine or describe the performative equivalence between a machine and a human being when casual patterns of ordinary speech permit us to say that a machine thinks, calculates, speaks, understands, and so on?

An interest in this question leads us to the old debate about what computers can and cannot do and to the question of whether the human brain functions like a computer. For our purposes here the battles between AI skeptics like Hubert Dreyfus or Joseph Weizenbaum and AI optimists like Alan Turing or Marvin Minsky can be left aside. What does matter, however, is to test a concept of information derived from the philosophy of language and oriented toward human action against these fundamental questions and to see if it has anything to say about them. For a criticism of anthropomorphic speech patterns applied to technical machines the task is therefore to ask whether an *explicitly articulated criterion of equal performance* can allow us to decide whether rhetorical humanizations of the machine express something true about them or whether they merely involve metaphors, which could then be classified as inappropriate or misleading.

Vehicles that decrease the human effort required to do difficult physical work—cranes for lifting, excavators for digging, flatbed trucks for carrying—can surpass specific human capacities to almost any degree. That fact stems above all from their specialized nature, which depends in turn on a well-nigh infinite restriction of the vast range of capacities shared by that most flexible of living beings, the human. This specialization engenders the efficiency of such vehicles, which produce "the same capacity" *in different ways*. So, for instance, in practice, machines have gears and use the law of levers to organize those gears into gear systems, whereas for living beings, wheels and gears play, in principle, no part. The free spinning of a wheel on an axle, or of an axle on its bearings, has nothing to do with the corporeal coherence that governs the way an organism's metabolism ensures the continual existence and replenishment of its components. We see other instances of these particularities, necessary for machines to do (otherwise) what humans do, when we compare the structures of vehicles that allow humans to fly or swim to the structures that allow animals to do so. The design principles governing flying machines (zeppelins, gliders, airplanes, helicopters) resemble in some cases those of natural flying beings, but the flying is accomplished in each case by completely different means. The same goes for, say, sharks and submarines.

And the same goes, again, for machines that *replace cognitive functions* rather than physical ones. When it comes to its specific capacities, even the simplest calculator surpasses human performance, if only at the price of specialization, and by virtue of structural principles that make its acts of calculating work differently than they do for humans, who manage things in their heads or with the aid of pen and paper.

The example of the calculator is, however, too narrow to get us to the *criterion of equal performance* we're looking for. In their case we are dealing with machines that aim to reproduce, performatively, the validity of arithmetic expressions or, in the case of more complex machines, to reproduce results that are fundamentally *calculable*. Calculations belong to more or less complex systems of rules for the generation of figures from figures. In other words, beginning with calculators as so-called *syntactical machines* can only be successful when you can express specific human capacities as calculations. For certain areas, this is a technically challenging task, which I take to be neither trivial nor unimportant. But they do not matter for the question of whether machines think or speak, or can perform equally well any other specific human cognitive functions, because discourse and acting, considered in their specifically human sense, are *fundamentally not calculable*. To establish a ground for these claims (which AI optimists would surely reject), we must begin the search for a criterion of equal performance with human language. In that case, the history of culture has given us, a millennium before the modern invention of calculating and speaking machines, an instance of the substitution of human speech by technology, namely, the written word. That history also evinces, conveniently enough, the philosophical dimensions of the concept of equal performance.

THE WRITING OF DISCOURSE

The fact that the daily practice of human beings affords no real alternatives to the principle of methodical ordering suggests how deeply rooted it is in our common and ordinary lives. For instance, it is indisputable that one must learn to speak if one wants then to learn to read and write. It is similarly clear that one does not learn to read solely by looking at books or written texts but rather through a combination of spoken words and gestures toward written signs, all of this under the aegis of an earlier mastery of spoken language. There are, in short, good reasons to speak of the *writing of spoken language,* even as we find it impossible to speak of the reverse, both in terms of the way a single person learns how to read and for the development of language and writing in the history of culture as a whole. These reasons lie above all in the embeddedness of discourse in action and in the earlier mentioned characteristics of participatory and communal action within the communicative field. (Someone for whom such philosophical reflections do not warm the heart may be more impressed at this point by the idea that in German the word for "language," *Sprache,* comes from the word for "speaking," *sprechen,* just as the English word *language* comes from the Latin word *lingua,* "tongue." Had things gone the other way around, we might call language *Schreibe,* that is, "writing," or, in English, *scriptage.*)

The writing of language itself is a *goal-oriented, rational* undertaking. When, as the product of a broad consensus, literacy initiatives were undertaken in the nations of the third world, they spread under the assumption that only the mastery of language in word and script, because it opens onto all the qualities of participation and self-determination stemming from the gains made in the Age of Enlightenment, can provide an adequate entry into the highest forms of modern culture, science, and the like. For scientific theories of communication and information, on the other hand, it suffices to emphasize the capacities of written language, to peel spoken language away from the ephemerality of the moment and the situation and make it moveable across space and time. This is especially apparent when it comes to early forms of writing, including letters, documents, laws, scripture, scientific treatises, and other culturally sacred forms.

It is (to make a slightly trivial methodical point) the case that someone can learn to read without learning to write (by hand or by machine). But it is also (methodically) the case that no one can learn to read if there is no writing at all. As a result, it will be important to divide a general sense of literacy into two components: learning to read and learning to write.

What does that distinction teach us? In the context of the myth of the naturalization of information, it returns us to the important difference between the natural and the cultural.

Already in Aristotle, we find a distinction between "natural" and "artificial" (which he refers to with the Greek work *technē*, "technical"). According to Aristotle, the natural is that which carries within itself the ground and the origin of its own characteristics and transformations. The technical, by contrast, is that which is produced by humans, ideally after some deep reflection regarding ends and means. Aristotle also tells us that this distinction is not an ontological disjunction, which would divide all being into the natural and the technical. It is rather a matter of *contrast between two aspects*. The same object, for instance, the marble statue in Aristotle's explanation of the four causes of being, has both natural and technical "characteristics," that is, it can be described under the aspect of its production (its formal cause, the shape of the statue, or its final cause, the reason why it has been made) and under the aspect of its material (the color or hardness of the marble).

Though Aristotle doesn't directly say so, we learn from his description of the technical as that which is planned and intentional to distinguish between an actor's *primary and secondary purposes*. The sculptor who makes the statue pays attention to any number of things, but probably not to the weight of the statue relative to the weight of the original marble block from which it is carved. Nonetheless, the weight of the statue is a "secondary technical effect" *(kata symbebekós)* of its production, even if no one intended its precise measure. Had the sculptor been interested instead in making a block of stone to measure bushels of grain, its weight would then be a primary purpose and its shape essentially a side effect.

All this will suffice for the moment as a minimal theory of the difference between the natural and the technical.

EQUAL PERFORMANCE AND METHODICAL ORDERING

The ending of the word *mechanization* (*Technisierung,* as we use it in "the mechanization of language") points to the fact that something is being artificially altered. It does not, however, necessarily suggest that whatever is being altered is itself something natural. The question of to what extent human language and communication are natural or artificial will therefore remain open here. (Evolutionary biologists who seek to develop a science of the human being have this problem to solve. But it cannot be done without some consideration of the natural–artificial distinction.) These days, in any case, we find ourselves not at some fictional cusp of the evolution of animals into human beings but in the twenty-first century, in which languages are unavoidably *cultural phenomena.* A glance at the etymology of the word *culture,* from the Latin *cultivare,* "to cultivate," reminds us that at its origin the concept refers to the intervention into and alteration of a pregiven nature. Let us not restrict the idea of culture, then, to the field of aesthetic activity (or even to the sciences or to political institutes) but recognize that it includes the planting of fruit trees, viniculture, the cultivation of bacteria (e.g., in yogurt), and so on. Humans intervene in the natural world as growers, shaping it to meet their various needs.

Whether language itself is to be regarded as a *technical* medium for the organization of communication and cooperation and for the management of human life matters little for what follows. All we need remember is the bare fact that "language" involves communication between people.

The first and most important form of the *mechanization (Technisierung) of discourse* happens with the invention of *writing.* Little is known about the historical process whereby this happened; the most concrete ideas we have appear in speculative philosophy (as in Plato) or literature (in Thomas Mann). We may as a kind of proxy for whatever historical events surround that invention consider the learning process of a single individual and the situation whereby language becomes mechanized in the individual. Everyone who learns to read and write learns them as distinct actions. And anyone who has learned those skills may draw from that learning some conclusions that are important to the myth of the naturalization of information:

- *Reading* is in some sense a *reversal of writing*. Only written things can be read. And no one can write who does not know how to read writing.

- *Reading and writing* take place *through human action*; writing is not possible without actions of movement and production. To produce a document, a writer moves in an embodied way according to a pattern of cultural gestures. Such gestures are not innate but learned. In terms of philosophy and action theory, kinesis (movement) and poiesis (production) are the grounds for the arts of writing (and, therefore, reading).

- The *writing* of spoken discourse *conserves* some particular features of the spoken word and *loses* others. When discourse moves into writing, it loses loudness and pitch, facial expressions and gestures, the implicit or artfully expressed signals of body language, the social role of the speaker, the situation, and much more. What remains, however, is that which the speaker wanted to share with the listener. Remembering Aristotle, we might say that the division here is between primary and secondary purposes.

- The spoken and written word *perform equally* in a number of ways. Schoolchildren learn this when they do dictation exercises in which they must write down what a teacher reads aloud to them. Writing preserves the *meaning* and *value* of the spoken words. In the standard case of a human writer (and in contrast to what goes on in a computerized dictation system) a sufficient understanding of the spoken words remains even when the words tell the story of something fictional or untrue. Different words can produce the same sounds; someone who does not speak a foreign language cannot transcribe it in dictation. The use of punctuation or the perils of orthography (whether something happening "at his bərθ" refers to a baby [birth] or a sailor [berth]) can only be navigated thanks to context. Semantics here methodically precedes value (via pragmatics).

- The technical medium that is the writing of spoken language serves clearly recognizable goals: whether one writes a letter or a diary entry, a research protocol or a legal document, one always intends the written text as a *tool for the transporting* of the spoken *through space and/or time*. In this way the forms of communication in spoken discourse spread to regions (and regions of life) in which the communication partner is not present, or not yet present, or does not have access to the structures of spoken language.

The mechanization *(Technisierung)* of language, understood as the writing of discourse via the human action of writing (and reading), is therefore a cultural technique with clearly determinable means and ends. It serves for the *equally performing substitution of the spoken word through writing, in order to transport the spoken over space and time.* The technical substitute performs equally well for certain specific aspects of language and not so well for others. The history of culture thus presents us with not only the most prominent example of the mechanization of language but the most instructive one, one that takes place long before the first forms of communications technology are developed and put into practice (already in ancient military and economic contexts). The mechanization of language and of communication is thus no child of communication technology, informatics, cybernetics, or any other similarly youthful science.

We are now in a position to understand in the broadest sense what counts as a performatively equal substitution for a human cognitive capacity. Whether it has to do with the control and regulation of the traffic system of a large urban area or the management of an enormous petrochemical corporation, whether we're talking about an automated system for processing passports that includes camera surveillance, a reading machine for the visually impaired, or a speech-recognition system for computers—in every single case, *performatively equal technological capacities substitute for primarily human cognitive ones.* Accordingly, the first step is to describe sufficiently the specific goal of the substitution, in order to grasp the *criterion for the function of the machine constructed to achieve it.* It is true without exception that information-processing machines can imitate highly specialized applications of human capacities. These capacities are themselves methodically irreducible. They are primary; as products of culture, they precede every mechanization.

This does not rule out the idea that mechanization has consequences that do appear (or did not appear) in nonmechanized practices. Think of the spread of mobile telephony or of the Internet. In the contemporary university the sending of letters by mail has almost entirely disappeared, except for certain documents that require signatures, seals, or stamps.

Philosophically, what matters is that this status of methodical primacy of human cultural activities lays out the framework within which the *limits of mechanical substitutability* operate. The development of criteria of equal performance, so critical to the work of the computer scientist or the engineer, requires as a matter of methodical primacy *having a sufficient explanation of the human cognitive capacities* themselves.

At this point we can return to the earlier critique of the deficits of hermeneutics, which will permit certain exemplary limits to emerge.

"SPEAKING" MACHINES

Just as, in the seventeenth and eighteenth centuries, mechanics and clockmaking reached new levels of perfection, the best practitioners of these arts began to cast about for tasks more difficult even than the construction of clocks (and the astronomical models that accompanied them). And so humanity arrives at the first golden age of robotics. The great triumphs of this period included a cymbal player and a puppet that could write a word on a piece of paper or do simple calculations; these led in turn to an increased fascination with such visible expressions of life as the blinking of eyelids and the movement of a breathing chest, whose results can still command wonder. Perhaps the greatest marvel of the age was the famous chess-playing automaton (built by Wolfgang von Kempelen in 1770), which toured the globe through the 1840s, though it was later revealed to be a hoax. In the dream of a chess-playing machine we may witness (as we do today in competitions between humans and computers) the limits of the mechanical arts and the superficial glimmer of something like equal performance.

From then until now is a long way. Nonetheless, the big question remains the same: are there limits to the mechanical substitutability of human capacities? Today this debate presses us in two directions: first, toward the unavoidable technical models of the brain provided by neuroscience, and second, toward the possible futures of research into, and development of, artificial intelligence and artificial life. (I leave aside a possible third field of discussion, the analytic philosophy of mind, since it tends merely to chase developments in the natural and technical sciences.)

For most experts in robotics it is obvious that any given robot aims to imitate a highly specific human capacity and to perform that capacity much faster or better than a human can. It is equally obvious that the replacement of humans by robots in a *diffuse and generalized* sense, so that all human functions could be replaced by robotic ones, is neither possible nor desirable. A shopping robot that had to find its way through city streets and buy clothes for its owner would soon run into technical limitations. More generally, this is a matter of the argument about independent existence. Positioning oneself in an environment is not just a matter of being oriented in space and time. It has to do with having a sense of one's own body, one's own intentions and interests, a structure of needs and a picture of the world. None of these is technically fungible.

Robotics is therefore not really a matter of science fiction. It is rather a field in which, for instance, the technologies of space exploration are discussed for good reason and with great seriousness: which capacities are unique to a human traveler; which can be replaced by remote control (telematics) or cannot be, if the signal's travel time is too long; and which ones can be automated? The major limits of automation lie in the *planning of functions designed to react to the unplanned or unexpected,* which come up against the logical boundaries of self-contradiction. Only humans can react *creatively* to something they've never seen before.

As far as the naturalization of information is concerned, our interest lies mainly with *linguistic capacities.* As a stand-in for a host of other problems, I will simply explain here why a machine cannot say "I." In scholarship we usually say that a machine cannot adopt a "first-person perspective." Among analytic philosophers of mind, this has produced a debate about the difference between first- and third-person perspective. Unfortunately, none of the participants seem to have taken account of the more general intellectual and cultural inheritances under which this debate takes place.

For instance, no one seems to have noticed that behind the idea of third-person perspective lies none other than the theory of descriptivism, which is purportedly the fundamental position of any serious scientist. As a result, it is easy to overlook the fact that *scientists can only act as authors of truth-claims in the first person.* It's also apparently easy to forget about second-person perspective, which plays no role in these debates. Why not?

I wrote earlier about the two ways in which the practice of ordinary life helps us act "as if" requests made by machines of humans, or of humans by machines, could be understood and responded to. It now seems possible to carry the example further. Imagine that among a group of (human) addressees, certain people would be called on by a machine (perhaps by the reciting of a previously assigned number): this amounts to an *equally performing substitution of the salutation "you."* To explain why requires a little detour through the ways that humans ordinarily use personal pronouns (and, for that matter, possessive ones).

To get there, let us begin by methodically reconstructing—not with an eye toward empirical developmental psychology but extending our interest in a philosophy of language grounded in theories of action—the process whereby a child is taught to use personal and possessive pronouns. From the perspective of a methodical reconstruction, what *goals cannot be reached without the use of those pronouns?*

I'll begin again with participatory and communal action involving mutual requests, which establish the initial steps for any cooperation. I'll also assume that the participants in a conversation have access to no specific linguistic forms of address and that only the direction of gaze or speech determines to whom any given person is speaking. Imagine that we come upon a situation in which we would like to ask one child to do something with or to another—to give the second child a toy, for example. In basic methodical situations we can use proper names. The personal pronouns in dative and accusative forms (Give her the ball! Leave him alone!) appear, as a next step, as *abbreviating variables, a shorthand, that refer contextually to a specific person and take the place of the proper name.*

Because these contexts involve the *attribution of actions* (from one person or another) *or the assignment of merit and blame,* the I- and you-forms serve their goals admirably. One asks of a certain (positive or negative) act, "Who did this?" Answers like "I" or "he" or protests like "You did!" can follow. In short, *personal pronouns are,* in contrast to situationally independent proper names, *situationally dependent attribution words.*[1] Some explanations of the word "self-consciousness" would require us at this point to climb the ladder of self-reference, arguing that all users of personal pronouns (including plural or grammatically declined ones) not only can but must know that their conversational counterpart also knows about personal pronouns, and that both of them know that they both know about personal pronouns, and so on. None of that is necessary here.

So you can set aside the question of reflexivity and the iteration of reflexivities (what analytic philosophers would call a "theory of mind"[2]) to observe that people in action and discourse communities make one another mutually responsible because *to make each other responsible* means to mutually attribute something to and with another. Proper names allow language to accomplish this task. Contrary to the extensive discussions in language philosophy, then, proper names are *not oriented toward predicative discourse* but intended in a context of cooperation, that is, request-oriented discourse. We need proper names to *address people* in situations where more than one person could potentially be addressed. A kindergarten teacher needs names to call children when those calls cannot be (or cannot be every single time) substantively replaced with a physical, facial, or gestural action or by the expedient of picking a child up and bringing him over. (This situational use of proper names opens up the possibility, quite obviously, of using those names in the absence of the people to whom they refer and thus independently from them.)

What matters here is the vital abbreviating function of everyday language, which replaces proper names with pronouns (literally, pro-names, things that stand in the place of the name) in situationally bound language contexts. (You see why analytic philosophy of language and philosophy of mind struggle with the problem of first- and third-person perspective: they do not primarily treat language as action and action as cooperation but, following the analogy of a theory of organisms, place the isolated individual in a hypothetical, natural environment. Beginning with the experience of the actual world of human life, in which a person can only survive infancy and become a member of the society of survivors if she has developed a minimal competence in the adjudication of social forms of merit and blame, allows us to avoid the problems created by such scientific hypotheticals.)

And so we reach a *fundamental limit for language-using (and thinking) machines.* Describing the limit cannot be as simple as asserting that machines cannot say "I" or "you"—obviously any properly designed machine can generate those words in sound or script. More pointedly, it seems clear that under certain technological conditions it may be possible to produce a technical indistinguishability between statements made by a person and a machine (as might happen in a modified Turing test), though at the cost of the depriving the human involved of her personhood.

As a first step, one might say that a machine cannot use personal or possessive pronouns *meaningfully*. But then you would have to define *meaningful* in such a way as to not lose the *attributional character* of the speech acts involved, neither on the side of their author nor on the side of the interlocutor(s) to whom something is attributed.

We do not assign blame or merit to machines. Machines may have natural characteristics, so to speak, that are not those intended by their inventors, builders, or users (who are ultimately responsible for the qualities they intend). But these characteristics are, in keeping with the Aristotelian lesson on coincidence, unintended side effects of human actions, understandable as secondary effects precisely because they could have been primary ones. (Aristotle gives the example of a man who goes to the market to meet a friend but "coincidentally" meets someone who owes him money; the man could also have gone to the market to meet his debtor. But if, while at the market, the man is caught in an earthquake, it is not the case that he could have gone to the market so that he could live through an earthquake. The earthquake is not "coincidental.")

When a construction crane blows over in a storm and hurts someone, we do not blame the wind or the crane for the disaster; we look to the machine's designer, manufacturer, or operator. That we assign (moral and legal) responsibility only to legal persons and not to things, forces of nature, animals, and only in a limited way to quasi-legal persons (children, the mentally disabled, or those suffering from a temporary debilitating illness) is a major achievement of contemporary culture. Recognizing that achievement gives us another reason why the linguistic means for the attribution of merit and blame cannot in principle be assigned to machines, and why, in other words, it would not be useful to do so. And this means, in turn, eliminating the possibility of their using possessive and personal pronouns, because the latter form a specific and important part of the practice of language.

This insight applies also to our understanding of the Turing test as well as to a number of other debates in research on artificial intelligence and in philosophy of mind, all of which remain trapped in the naturalization of information and the myths that bring it to life.

Models of the Natural

Once humanity recognizes itself (and its peers) as being methodically primary in the realms of action and discourse, it becomes possible to think of technology as methodically secondary, in two senses: on one hand, intellectually, as know-how and means, and on the other, as object and site of responsibility. This does not require us to idealize the situation; indeed, we cannot overlook the fact that the predictability of primary and secondary consequences has its limits. Artifacts have the characteristic of outlasting their makers and users. Someone who encounters an object need pay no attention to its original purpose but can reinterpret it and bend it toward new possibilities. This reinterpretability is in fact the fundamental principle of technological civilization; the historical development that leads from the wagon wheel to complex gearing systems illustrates it clearly enough.

Even so, a methodically secondary technology is nonetheless and in its own way *primary for the engagement with the natural world*. Regardless of the possibilities and richness of aesthetic, emotional, or moral dealings with nature, reasonable engagement with nature is primarily and inevitably interventionist. Pace Goethe's objections to Newton's splitting of light in the prism, pace even Galileo's misunderstanding of his own experiments, it is not just the scientific researcher in the laboratory who dominates nature with his apparatus. Prehistoric hunters and gatherers, and especially the early breeders of plants and animals, the planters of fields and forests, all intervened in nature in order to live. And they acquired their shareable know-how by experience, on the basis of functioning and nonfunctioning, the success and failure of their interventions. Nature allows us only to understand it through technical models.

Just as, then, one can model natural processes using mechanical systems (in the widest sense, including electrodynamic, thermodynamic, or chemical ones), so one can model those processes using machines that replicate and substitute for human cognitive capacities.

There would be, for instance, nothing wrong with a successful modeling of the brain as the telephone network of a major metropolis.

But one knows nonetheless that such a model would be completely inadequate. It would only be culturally and historically noteworthy insofar as it would repeat once again the gesture of all previous models of the central nervous system, which borrow each time from the most modern technological possibilities of their era. This applies as much to Descartes, whose substance-oriented model of the world of the *res extensa,* in which natural events occur only as the result of the propagation of pressure and impacts, reappears in his figuration of conduction in nerve canals, as it does to contemporary theories of "neural networks."

The all-too-common confusions that plague contemporary naturalism and its subfields (empiricism, epistemological realism, and scientific formalism) all begin when the modeling of nature by artifacts forgets that the artifacts and their functions themselves express, and are determined by, the achievements and demands of human culture.

SIX
CONSEQUENCES

What changes in the world . . .

- when we see that the naturalization of information is nothing but a myth, one that stems from the investments of (logical empiricist) philosophy?
- when the state of the debate changes so that naturalism is no longer our only intellectual alternative and can be replaced by a methodical culturalist theory of communication and information?
- when this (constructive) alternative covers all the technical accomplishments of the naturalistic approach, while avoiding all of its philosophical errors (e.g., with regard to the assessment of the abilities of machines)?
- when a collaboration among the natural, technical, formal, and human sciences comes into view?

No one should or can be taught that thinking of the world in her own language is irrelevant to (supposedly) language-independent necessities or experiences. This is not a matter of improving the planet but of enlightenment, in the sense of a critique of false philosophies. No one who wishes to be a successful computer programmer or neuro-scientist or analytic philosopher need trouble herself with the critique of naturalism in order to do so. In fact, belonging to the philosophical mainstream makes, socially and psychologically speaking, for an especially effective pathway to success.

But there is *one* problem that the naturalist cannot escape, or resolve: he cannot see his naturalism from the inside, from the perspective of its own implementation. Like the machine, he cannot say "I." And so the naturalist picture of the human, so dominant today in both philosophy and the natural sciences, takes on a schizophrenic relation to morality and the law.

The aim of this book has been, therefore, minimally to produce another mode of speech, minimally to allow a rethinking of the philosophical grounds of the naturalist position. Its success depends on its not including decisive errors and, of course, on its finding here and there a few readers. In any case, no one can any longer say that there has never been an alternative to naturalism, or to the naturalization of information, or claim that the rationality of information rests, in the fully naturalized sciences, in the *only* place that it might call home.

NOTES

TRANSLATORS' INTRODUCTION

1. See, e.g., Alex Wright's *Glut: Mastering Information through the Ages* (Ithaca, N.Y.: Cornell University Press, 2008); James Gleick's *The Information: A History, a Theory, a Flood* (New York: Vintage, 2012), or John Durham Peters's *The Marvelous Clouds: Toward a Philosophy of Elemental Media* (Chicago: University of Chicago Press, 2015).

2. Michael E. Hobart and Zachary S. Schiffman, *Information Ages: Literacy, Numeracy, and the Computer Revolution* (Baltimore: Johns Hopkins University Press, 1998); Bernard Geoghegan, "From Information Theory to French Theory: Jakobson, Lévi-Strauss, and the Cybernetic Apparatus," *Critical Inquiry* 38 (Autumn 2011): 96–126.

3. Milman Parry, *The Making of Homeric Verse: The Collected Papers of Milman Parry* (Oxford: Oxford University Press, 1987); Albert Lord, *The Singer of Tales*, 2nd ed., ed. Stephen Mitchell and Gregory Nagy (Cambridge, Mass.: Harvard University Press, 2000).

4. See the first chapters of Gleick, *The Information*.

5. Peter Janich, *Protophysics of Time: Constructive Foundation and History of Time Measurement* (Dordrecht, Netherlands: D. Reidel, 1985), xxiv.

6. Ibid., 83.

7. Dirk Hartmann and Peter Janich, eds., *Methodischer Kulturalismus:*

Zwischen Naturalismus und Postmoderne (Frankfurt am Main, Germany: Suhrkamp, 1996), 68.

8. Peter Janich, "Methodical Constructivism," in *Issues and Images in the Philosophy of Science: Scientific and Philosophical Essays in Honour of Azarya Polikarov,* ed. Dimitri Ginev and Robert S. Cohen, 181–82. Berlin: Kluwer Academic, 1997.

9. Garfinkel, *Toward a Sociological Theory of Information,* ed. Ann Warfield Rawls (Boulder, Colo.: Paradigm, 2008).

10. Janich, "Methodical Constructivism," 187.

11. See the introduction to Dipesh Chakrabarty's, *Provincializing Europe* (Princeton, N.J.: Princeton University Press, 2008), where, after noting that "one empirically knows of no society in which humans have existed without gods or spirits accompanying them," Chakrabarty goes on to say that he takes "gods and spirits to be existentially coeval with the human" and therefore makes no use of sociology of religion (28).

12. Here a useful intertext would be Derrida's critique of speech-act theory in *Limited Inc.,* trans. Jeffrey Mehlman and Samuel Webber (Evanston, Ill.: Northwestern University Press, 1988).

13. Peter Janich, "Technology and the Levels of Culture," *Poiesis and Praxis* 1, no. 4 (2003): 267.

14. Such a naturalistic case, Janich argues in *Der Mensch und andere Tiere: Das zweideutige Erbe Darwins* (Frankfurt am Main, Germany: Suhrkamp, 2010), fails to acknowledge that humans and animals exist as both *Naturgegenstände* (natural objects or facts) and products of a *Kulturgeschichte* (cultural history). Describing biological differences between humans and animals is not the same as ascribing to them differences that are products of cultural practice and habit (including moral and legal distinctions).

15. It is worth noting that, as with all things, the distinction between human and nonhuman animals (or indeed any distinction among animals at all) will simply depend on which criteria constitute the applicable framework for comparison and thus establish the line that separates difference in kind from difference in degree. Derrida's critique of the long history of the *animot* and of the various historical claims made for the specific, kind-difference-creating quality of human rationality offers one analysis of the operations of these categories. Jacques Derrida, *The Animal That Therefore I Am,* ed. Marie-Louise Mallet, trans. David Wills (New York: Fordham University Press, 2008). But it is also worth noting that large-scale differences among animals may nonetheless be captured by large enough struc-

tures of similarity (the desire to flourish, for instance, or the capacity to feel pain) so as to justify a full-blown program of animal rights, without necessarily producing the kind of biologistic reduction that would simply be the obverse side of the Derridian *animot*, as one sees in Sue Donaldson and Will Kymlicka, *Zoopolis: A Political Theory of Animal Rights* (New York: Oxford University Press, 2013).

16. The problems posed by the insistence on rationality as a marker of kind-difference for the human are extensively discussed in Donaldson and Kymlicka, *Zoopolis*, as well as in Martha Nussbaum, *Frontiers of Justice: Disability, Nationality, Species Membership* (Cambridge, Mass.: Belknap Press of Harvard University Press, 2007).

17. http://www.duden.de/rechtschreibung/Banause.

1. INFORMATION AND MYTH

Janich's original text offers limited bibliographic data and includes no notes. The translators have moved all of his bibliographic information into notes, expanding it and referring to English originals or translations where appropriate.

1. Janich uses *Legende* throughout. The connotations of the English cognate *legend* not having quite the same range as the German, we have often translated it as "myth," with an ear for the latter's echoes with the work of Roland Barthes. —Trans.

2. Elmar Seebold, ed., *Etymologisches Wörterbuch der Deutschen Sprache*, 22nd ed. (Berlin: Walter de Gruyter, 1989). Orinally edited by Friedrich Kluge.

3. Adolf Hanle, *Meyers Enzyklopädisches Lexicon*, vol. 14 (Leipzig, Germany: Bibliographisches Institut, 1975).

4. See Sascha Ott, *Information: Zur Genese und Anwendung eines Begriffs* (Konstanz, Germany: UVK, 2004). (In English, see James Gleick, *The Information: A History, a Theory, a Flood* [New York: Vintage, 2012], or, for something more philosophical, Luciana Floridi's *Information: A Very Short Introduction* [Oxford: Oxford University Press, 2010]. —Trans.)

5. Norbert Wiener, *Cybernetics: or, Control and Communication in the Animal and the Machine*, 2nd ed. (Cambridge, Mass.: MIT Press, 1965), 132.

6. Janich uses this opposition between object language *(Objektsprache)* and metalanguage *(Metasprache)* to distinguish statements made within the framework of a scientific discipline (e.g., statements by physicists about the nature of gravity) and the discourse that surrounds those statements and justifies them (language about the validity of the experimental method). Establishing strict distinctions

between what he will later call "levels" of language about things constitutes a central feature of his philosophical method. —Trans.

7. The section "Arguments Proving the Existence of God and the Distinction between the Soul and the Body" appears as part of Descartes's
responses to the second set of objections to his *Meditations on First Philosophy.* It can be found in René Descartes, *The Philosophical Writings
of Descartes,* trans. John Cottingham, Robert Stoothoff, and Dugald
Murdoch (Cambridge: Cambridge University Press, 1984), 2:113–20.
—Trans.

8. Shannon published "A Mathematical Theory of Communication" in
two parts in *The Bell System Technical Journal* 27, no. 3 (1948): 379–423
and 27, no. 4 (1948): 623–56. The articles were republished, along with
a lengthy introductory essay from Weaver, as part of a small book,
The Mathematical Theory of Communication (Champaign: University of
Illinois Press, 1949).

2. LEGACIES

1. Heinrich Hertz, *The Principles of Mechanics Presented in a New Form,*
trans. D. E. Jones and J. T. Walley (New York: Dover, 1956), 1, translation modified.

2. Ibid.

3. Lorenz (1903–89), winner of the 1973 Nobel Prize in Physiology or
Medicine, was an Austrian zoologist and ethnologist perhaps best
known for his studies of imprinting in animals and other studies that
regarded animal behavior as an evolutionarily adaptive trait. Vollmer
(b. 1943), a physicist and philosopher, wrote *Evolutionäre Erkenntnistheorie* [Evolutionary epistemology], 8th ed. (Leipzig, Germany:
S. Hirzel, 2002), among other books. —Trans.

4. Hertz, *Principles of Mechanics,* 1.

5. Antagonistic to both neo-Kantianism and Husserlian phenomenol-
ogy, the Vienna Circle (Wiener Kreis) was a group of philosophers,
scientists, logicians, and mathematicians who met regularly at the University of Vienna from 1924 to 1936. Organized primarily by Moritz
Schlick, who began the meetings in his home in the 1920s, the group
included Otto Neurath, Rudolf Carnap, Kurt Gödel, and other major
figures; it solicited visits and conversation with such thinkers as Carl
Gustav Hempel, Alfred Tarski, Bertrand Russell, Willard Van Orman
Quine, Karl Popper, and Ludwig Wittgenstein. Its major philosophical positions, which can be gathered loosely around the name "logical
empiricism," attempted to put philosophy on the same ground as the

natural sciences by adopting strict rules regarding epistemological legitimacy. Those rules could involve a "reduction" of mathematical or syntactic statements to logic (as in the work of the early Wittgenstein, Russell, or Gottlob Frege), an attempt to develop a universally valid language (Carnap), or a more general insistence that only "verifiable" statements can be scientifically meaningful. Much of the group's influence came after 1945, when its ideas, carried into the United States by wartime refugees like Hempel and Carnap, profoundly shaped the development of analytic philosophy and the philosophy of science. —Trans.

6. The trilemma, which refers to the Baron Münchhausen story of pulling himself up by his own hair, describes a thought experiment in epistemology. The trilemma consists of responses to the question "How do I know that X is true?" which, as Karl Popper had noted, could only be of three types: circular, regressive, and axiomatic, or, as Albert put them, a logical circle, infinite regression, and a "break in searching at a certain point." Hans Albert, *Traktat über kritische Vernunft* [Treatise on critical reason] (Leipzig: J. C. B. Mohr, 1991), 15. —Trans.

7. Albert Einstein, *Geometrie und Erfahrung* (Berlin: Springer, 1921), 4; available in English online at http://pascal.iseg.utl.pt/~ncrato/Math/Einstein.htm, where the cited sentence appears in the third paragraph. —Trans.

8. Rudolf Carnap, *An Introduction to the Philosophy of Science,* ed. Martin Gardner (New York: Basic Books, 1966), 183.

3. ARTICLES OF FAITH

1. The first essay appeared in *International Encyclopedia of Unified Science* 1, no. 2 (1938), the second in *Erkenntnis* 8 (1939). Both are reprinted in Charles Morris, *Writings on the General Theory of Signs* (Berlin: De Gruyter Mouton, 1971), cited here. Janich is citing from the translation into German, *Grundlagen der Zeichentheorie: Ästhetik der Zeichentheorie* (Munich: Carl Hanser, 1972). —Trans.

2. W. V. O. Quine, "Epistemology Naturalized," in *Ontological Relativity and Other Essays,* 69–90 (New York: Columbia University Press, 1969).

3. *Rückkopplung* is "feedback" in the technical sense rather than the feedback *(Rückmeldung)* someone gives someone about a project or idea. —Trans.

4. The word *Reflexionsterminus* has a special meaning in Janich's work. It refers to the ways in which a certain casual shorthand can

produce the illusion of substance where there is none, abstracting ontology from process. For instance, consider the word *life*. One speaks of "life" if one wants to address a particular aspect of living things, namely, the fact of their being alive. But biology does not require a new concept of "life" to address the inadequacies of "living"; "life" is not a separate subject for the field but only an abbreviated way of speaking about living things. For more, see Janich, "Reflexionsterminus," in *Enzyklopädie Philosophie und Wissenschaftstheorie*, ed. Jürgen Mittelstraß, vol. 3 (Stuttgart, Germany: Metzler, 1995). —Trans.

5. *Scientific American*, September 1971, 80.

6. On negentropy, see chapter 14 of Brillouin's *Science and Information Theory* (New York: Academic Press, 1956), now available in the public domain at HathiTrust, https://babel.hathitrust.org/cgi/pt?id=mdp.39015006410149. —Trans.

7. Holger Lyre, *Informationstheorie: Eine philosophisch-naturwissenschaftliche Einführung* [Information theory: A philosophical and scientific introduction] (Stuttgart, Germany: Fink, 2002), 48.

8. Cited ibid., 47.

4. INFORMATION CONCEPTS TODAY

1. A "Markov process" (named after the Russian mathematician Alexey Markov [1856–1922]) involves a stochastic (statistically indeterminate) set of variables that describe the state of a system at a given moment, as long as such a system has the "Markov property." A system has the Markov property if the future probabilistic state of the system depends only on the current state of the system and not on its past. For this reason, systems with the Markov property can be called "memoryless." Shannon discusses stochastic Markov processes in *Mathematical Theory of Communication*, 45–48. —Trans.

2. Morris, "Esthetics of the Theory of Signs," 416.

3. These two pairs are near-synonyms and would usually both be translated into English as "success" and "failure." *Gelingen* and *Mißlingen* have more of a sense of action or process, of accomplishment or failure to accomplish or carry out some act. Janich uses the two pairs together throughout to emphasize different aspects of communicative activity; here we translate *Gelingen* and *Mißlingen* as "functioning" and "nonfunctioning" and *Erfolg* and *Mißerfolg* as "success" and "failure." —Trans.

4. In English the word is *podiatrist*, but compare *pedology*, the study of soils, from the same Greek root. —Trans.

5. Pier Lucio Anelli et al., "Molecular Meccano. 1. [2] Rotaxanes and a [2] Catenane Made to Order," *Journal of the American Chemical Society* 114 (1992): 193.

6. Janich paraphrases Goethe's *Faust,* when the angels say that "whoever strives with all his might, him we can save" (11936–37). —Trans.

7. Monroe W. Strickberger, *Genetik* (Leipzig, Germany: Fachbuchverlag, 1988).

8. Werner Ebeling and Rainer Faistel, *Physik der Selbstorganisation und Evolution* (Berlin: Akademie, 1986), 304.

9. Manfred Eigen, "The Origin of Biological Information," in *The Physicist's Conception of Nature,* ed. Jagdish Mehra, 594–633 (Dordrecht, Netherlands: D. Reidel, 1973).

10. Ebeling and Faistel, *Physik,* 312; Friedhart Klix, *Information und Verhalten: Kybernetische Aspekte der organismischen Informationsverarbeitung* (Bern, Switzerland: Deutscher Verlag der Wissenschaften, 1976).

11. See, e.g., Konrad Lorenz, *The Foundations of Ethology* (New York: Springer, 1981). —Trans.

12. See Helmut Schnelle, "Information," in *Historisches Wörterbuch der Philosophie,* ed. Joachim Ritter and Karlfried Gründer (Basel, Switzerland: Schwabe Basel, 1976), 4:355.

13. See Wolfgang F. Gutmann and Michael Weingarten, "Die biotheoretischen Mängel der evolutionären Erkenntnistheorie," *Zeitschrift für allgemeine Wissenschaftstheorie* 21, no. 2 (1990): 309–28.

14. See Joachim W. Engels, "Gene, die uns bewegen: Von der Definition der Gene zur Sequenzierung des menschlichen Genoms," *Sitzungsberichte der Wissenschaftlichen Gesellschaft an der Johann Wolfgang Goethe Universität* 36, no. 2 (Stuttgart, Germany: Franz Steiner, 1998), 115.

15. Edward O. Wilson, *Consilience: The Unity of Knowledge* (New York: Vintage, 1989), 291. The German translation that Janich uses abandons the word "consilience" (even in its title, *Die Einheit des Wissens* [The unity of knowledge] [Munich, Germany: Siedler, 1998]), and so it does not appear in his citations in German. Wilson describes "consilience," literally a "jumping together" of knowledge, as "the key" to the "unification" of the natural sciences, social sciences, and humanities in a single field (7). —Trans.

16. Though E.O. Wilson has won many prizes, he has never won a Nobel.

17. Konrad Lorenz, *Behind the Mirror: A Search for the Natural History of Human Knowledge,* trans. Ronald Taylor (New York: Methuen, 1977), 244–45, translation modified.

18. Bernd-Olaf Küppers, *Der Ursprung biologischer Information: Zur Naturphilosophie der Lebensentstehung* (Munich, Germany: Piper, 2000), 50.

19. "Das Geistlose in der Maschine"; Janich is punning on the phrase "the ghost in the machine," which first appeared in Gilbert Ryle's 1949 *The Concept of Mind,* where it parsed Descartes's mind–body dualism. The phrase reappears in the title of Alfred Koestler's *The Ghost in the Machine* (1969) and is in generic use to refer to the possibilities of artificial intelligence. *Geist* in German is "ghost" or "spirit," and *geistlos* is an adjective meaning "insipid," "unimaginative," or, as here, "spiritless." —Trans.

20. Janich consistently uses "know-how" in English in the original text. —Trans.

21. Hellmut Willke, *Systemtheorie I: Grundlagen: Eine Einführung in die Grundprobleme der Theorie sozialer Systeme,* 3rd ed. (Stuttgart, Germany: UTB, 1991), 8.

22. Achim Stephan, *Emergenz: Von der Unvorhersagbarkeit zur Selbstorganisation* (Muenster, Germany: Mentis, 1999), 14.

23. R. W. Sellars, *The Principles and Problems of Philosophy* (New York: Macmillan, 1926), 210.

24. In Janich this phrase ("hier feiert die Sprache") is in quotation marks. The relevant passage is from *Philosophical Investigations*: "Denn die philosophischen Probleme entstehen, wenn die Sprache *feiert*" (For philosophical problems arise when language *goes on holiday*). Ludwig Wittgenstein, *Philosophical Investigations,* 3rd bilingual ed., trans. G. E. M. Anscombe (Oxford: Blackwell, 2001), emphasis original. —Trans.

25. In English in the original. —Trans.

26. Wilhelm Windelband, "Rectorial Address, Strasbourg, 1894," *History and Theory* 19, no. 2 (1980): 169–85.

27. In German, available via Wikisource in a reproduction of the publication of the fourth edition by Friedrich Vieweg und Sohn (Braunschweig, 1929), https://de.wikisource.org/wiki/Die_Entstehung_der_Kontinente_und_Ozeane; in English, *The Origin of Continents and Oceans,* trans. John Biram (New York: Dover Press, 1966). —Trans.

28. C. P. Snow, *The Two Cultures and the Scientific Revolution* (Cambridge: Cambridge University Press, 1961).

29. Dietrich Schwanitz, *Bildung: Alles, Was Man wissen muß,* 7th ed. (Munich, Germany: Goldman, 2002); Ernst Peter Fischer, *Die andere Bildung: Was Man von den Naturwissenschaften wissen sollte* (Berlin: Ullstein, 2003).

30. "Operationalization" refers here to the ideas developed by the physicist Percy Williams Bridgman (1882–1961) in *The Logic of*

Modern Physics (New York: Macmillan, 1927). Bridgman argued that physical concepts like "time" or "length" ought to be defined not in terms of their properties (that is, metaphysically) but in terms of the physical or mental operations by which they are measured or described; thus "length" would mean nothing more than the set of operations whereby it emerges as an object of consciousness. Bridgman's work briefly influenced members of the Vienna Circle but was eventually repudiated by philosophers like Carnap and Hempel. Its major influence today is in the field of psychology. For a short critique of Bridgman, see Janich's *Protophysics of Time,* 34–35. —Trans.

5. METHODICAL REPAIR WORK

1. Such words are known in linguistics as "deictics" or "shifters" and in philosophy of language (which takes the word from the work of C. S. Peirce) as "indexicals." For a useful summary, see the chapter on deixis in Stephen Levenson's *Pragmatics* (Cambridge: Cambridge University Press, 1983). —Trans.

2. In English in the original. —Trans.

PETER JANICH
A Partial Bibliography

BOOKS IN ENGLISH

Brune, H., H. Ernst, A. Grunwald, W. Grünwald, H. Hofmann, H. Krug, P. Janich, M. Mayor, W. Rathgeber, G. Schmid, U. Simon, V. Vogel, and D. Wyrwa. *Nanotechnology: Assessment and Perspectives.* Berlin: Springer, 2006.

Janich, Peter. *Euclid's Heritage: Is Space Three-Dimensional?* Edited by Robert E. Butts. Vol. 52. Dordrecht, Netherlands: Springer, 1992.

———. *Protophysics of Time: Constructive Foundation and History of Time Measurement.* Edited by Robert S. Cohen and Marx W. Wartofsky. Boston Studies in the Philosophy of Science 30. Dordrecht, Netherlands: Springer, 1985.

ARTICLES IN ENGLISH

Janich, Peter. "Between Innovative Forms of Technology and Human Autonomy: Possibilities and Limitations of the Technical Substitution of Human Work." In *Robo- and Informationethics: Some Fundamentals,* edited by Michael P. Decker and Matthias Gutmann, 211–30. Berlin: Lit, 2012.

———. "Between Natural Disposition and Cultural Masterment of Life." In *On Human Nature: Anthropological, Biological, and Philosophical Foundations*, edited by Armin Grunwald, Matthias Gutmann, and Eva M. Neumann-Held, 95–110. Berlin: Springer, 2002.

———. "Commentary on 'Protophysics of Time and the Principle of Relativity.'" In *Physical Sciences and History of Physics*, edited by Robert S. Cohen and Marx W. Wartofsky, 191–98. Boston Studies in the Philosophy of Science 82. Dordrecht, Netherlands: Springer, 1983.

———. "The Concept of Mass." In *Constructivism and Science: Essays in Recent German Philosophy*, edited by Robert E. Butts and James Robert Brown, 145–62. London: Springer, 1989.

———. "Does Biology Need a Relativistic Revision?" *International Studies in the Philosophy of Science* 3, no. 2 (1989): 190–98.

———. "From Constructivism to Culturalism." In *Movement of Constructive Realism*, edited by Thomas Slunecko, 39–62. West Lafayette, Ind.: Purdue University Press, 1997.

———. "Health and Quality of Life: A Conceptual Proposal from the Perspective of Methodical Culturalism." In *Health and Quality of Life: Philosophical, Medical, and Cultural Aspects*, edited by Antje Gimmler, Christian Lenk, and Gerhard Aumüller, 47–59. Muenster, Germany: Lit, 2002.

———. "Human Nature and Neurosciences: A Methodical Cultural Criticism of Naturalism in the Neurosciences." *Poiesis and Praxis* 2 (2003): 29–40.

———. "Intuition, Language, and Action: Epistemological Notes on Intuition in Surgery." *Theoretical Surgery* 6, no. 2 (1991): 104–6.

———. "Methodical Constructivism." In *Issues and Images in the Philosophy of Science*, edited by Dimitri Ginev and Robert S. Cohen, 173–90. Berlin: Kluwer Academic, 1997.

———. "The Normative Foundation of Physics." *International Studies in the Philosophy of Science* 1, no. 2 (1987): 251–62.

———. "Performance and Description Perspective and the Problem of Scientific Transsubjectivity." In *Science, Medicine, and Culture: Festschrift for Fritz G. Wallner*, edited by Martin J. Jandl and Kurt Greiner, 27–37. Frankfurt, Germany: Peter Lang, 2005.

———. "Physics, Natural Science or Technology?" In *The Dynamics of Science and Technology: Social Values, Technical Norms, and Scientific Criteria in the Development of Knowledge*, edited by Wolfgang Krohn, Edward T. Layton Jr., and Peter Weingart, 3–27. Dordrecht, Netherlands: Springer, 1978.

———. "Physiology and Language: Epistemological Questions about Scientific Theories of Perception." In *Thermoreception and Temperature Regulation*, edited by J. Bligh and K. Voigt, 151–63. Berlin: Springer, 2011.

———. Preface to *Thought Experiment in the Natural Sciences*, by Marco Buzzoni, 9–11. Wuerzburg, Germany: Königshausen und Neumann, 2008.

———. "Technology and Levels of Culture." *Poiesis and Praxis* 1 (2003): 263.

———. "Truth as Success of Action." In *Scientific Knowledge Socialized*, edited by Imre Hronszky, Márta Feher, and Balázs Dajka, 313–26. Berlin: Kluwer Academic, 1988.

———. "Where Does Biology Get Its Objects From?" In *Organisms, Genes, and Evolution: Evolutionary Theory at the Crossroads*, edited by Dieter Stefan Peters and Michael Weingarten, 9–16. Stuttgart, Germany: Franz Steiner, 2000.

Janich, Peter, and M. Gutmann. "Species as Cultural Kinds: Towards a Culturalist Theory of Rational Taxonomy." *Theory in Biosciences* 117 (1998): 237–38.

BOOKS IN GERMAN

Janich, Peter. *Das Maß der Dinge: Protophysik von Raum, Zeit und Materie.* Frankfurt am Main, Germany: Suhrkamp, 1997.

———. *Der Mensch und andere Tiere: Das zweideutige Erbe Darwins.* Frankfurt am Main, Germany: Suhrkamp, 2010.

———. *Die Protophysik der Zeit: Konstruktive Begründung und Geschichte der Zeitmessung.* Frankfurt am Main, Germany: Suhrkamp, 1980.

———. *Emergenz—Lückenbüssergottheit für Natur- und Geisteswissenschaften: Ergänzt um eine Korrespondenz mit Hans-Rainer-Duncker.* Vol. 49, no. 2. Stuttgart, Germany: Franz Steiner, 2011.

———. *Euklids Erbe: Ist der Raum dreidimensional?* Munich, Germany: C. H. Beck, 1989.

———. *Grenzen der Naturwissenschaft: Erkennen als Handeln.* Munich, Germany: C. H. Beck, 1992.

———. *Handwerk und Mundwerk: Über das Herstellen von Wissen.* Munich, Germany: C. H. Beck, 2015.

———. *Kein neues Menschenbild: Zur Sprache der Hirnforschung.* Frankfurt am Main, Germany: Suhrkamp, 2009.

———. *Kleine Philosophie der Naturwissenschaften.* Munich, Germany: C. H. Beck, 1997.

———. *Konstruktivismus und Naturerkenntnis: Auf dem Weg zum Kulturalismus*. Frankfurt am Main, Germany: Suhrkamp, 1996.

———. *Kultur und Methode: Philosophie in einer wissenschaftlich geprägten Welt*. Frankfurt am Main, Germany: Suhrkamp, 2006.

———. *Logisch-pragmatische Propädeutik: Ein Grundkurs im philosophischen Reflektieren*. Weilerswist, Germany: Velbrück, 2001.

———. *Mensch und Natur: Zur Revision eines Verhältnisses im Blick auf die Wissenschaften*. Vol. 40, no. 2. Frankfurt am Main, Germany: Franz Steiner, 2002.

———. *Mundwerk ohne Handwerk? Ein vergessenes Rationalitätsprinzip und die geistesgeschichtlichen Folgen*. Vol. 53, no. 2. Stuttgart, Germany: Franz Steiner, 2016.

———. *Sprache und Methode: Eine Einführung in philosophische Reflexion*. Tuebingen, Germany: UTB, 2014.

———. *Was ist Erkenntnis: Eine philosphische Einführung (Beck'sche Reihe)*. Munich, Germany: C. H. Beck, 2000.

———. *Was ist Information? Kritik einer Legende*. Frankfurt am Main, Germany: Suhrkamp, 2006.

———. *Was ist Wahrheit? Eine philosophische Einführung*. Munich, Germany: C. H. Beck, 1996.

Janich, Peter, and Rolf Oerter. *Der Mensch zwischen Natur und Kultur*. Philosophie und Psychologie im Dialog 11. Goettingen, Germany: Vandehoeck und Rupprecht, 2012.

Janich, Peter, and M. Weingarten. *Wissenschaftstheorie der Biologie: Methodische Wissenschaftstheorie und die Begründung der Wissenschaften*. Paderborn, Germany: Fink, 1999.

EDITED BOOKS IN GERMAN

Janich, Peter, ed. *Der Mensch und seine Tiere: Mensch-Tier-Verhältnisse im Spiegel der Wissenschaften*. Stuttgart, Germany: Franz Steiner, 2014.

———, ed. *Entwicklungen der Methodischen*. Frankfurt am Main, Germany: Suhrkamp, 1992.

———, ed. *Humane Orientierungswissenschaft: Was leisten verschiedene Wissenschaftskulturen für das Verständnis der Lebenswelt?* Wuerzburg, Germany: Königshausen und Neumann, 2008.

———, ed. *Methodische Philosophie: Beiträge zum Begründungsproblem der exakten Wissenschaften in Auseinandersetzung*. Mannheim, Germany: Bibliographisches Institut, 1984.

———, ed. *Naturalismus und Menschenbild*. Vol. 1 of *Deutsches Jahrbuch Philosophie*. Hamburg, Germany: Felix Meiner, 2008.

———, ed. *Wechselwirkungen: Zum Verhältnis von Kulturalismus, Phäno-menologie und Methode*. Wuerzburg, Germany: Königshausen und Neumann, 1999.

———, ed. *Wissenschaft und Leben: Philosophische Begründungsprobleme in Auseinandersetzung*. Bielefeld, Germany: Transcript, 2006.

———, ed. *Wissenschaftstheorie und Wissenschaftsforschung*. Munich, Germany: C. H. Beck, 1981.

Janich, Peter, Mathias Gutmann, and Kathrin Prieß, eds. *Biodiversität: Wissenschaftliche Grundlagen und gesellschaftliche Relevanz*. Berlin: Springer, 2002.

Janich, Peter, and Dirk Hartmann, eds. *Die kulturalistische Wende: Zur Orientierung des philosophischen Selbstverständnisses*. Frankfurt am Main, Germany: Suhrkamp, 1998.

———, eds. *Methodischer Kulturalismus: Zwischen Naturalismus und Post-moderne*. Frankfurt am Main, Germany: Suhrkamp, 1996.

Janich, Peter, and Nikolaos Psarros, eds. *Der Autonomie der Chemie*. Vol. 3 of *Erlenmeyer-Kolloquium für Philosophie der Chemie*. Wuerzburg, Germany: Königshausen und Neumann, 1998.

———, eds. *Die Sprache der Chemie*. Vol. 2 of *Erlenmeyer-Kolloquium zur Philosophie der Chemie*. Wuerzburg, Germany: Königshausen und Neumann, 1996.

———, eds. *Philosophische Perspektiven der Chemie*. Vol. 1 of *Erlenmeyer-Kolloquium zur Philosophie der Chemie*. Mannheim, Germany: Spektrum Akademic, 1994.

Janich, Peter, and Christoph Rüchardt, eds. *Natürlich, technisch, chemisch: Verhältnisse zur Natur am Beispiel der Chemie*. Berlin: De Gruyter, 1996.

Janich, Peter, Peter C. Thieme, and Nikos Psarros, eds. *Chemische Grenzwerte: Eine Standortbestimmung von Chemikern, Juristen, Soziologen und Philosophen*. Weinheim, Germany: Wiley, 1999.

INDEX

(continued from page ii)

PETER JANICH (1942–2016) was a German philosopher of science. He wrote thirty books and more than two hundred articles. After earning his PhD in 1969 (philosophy, University of Erlangen), he taught at the University of Konstanz and at the University of Marburg until his retirement in 2007. He held research fellowships or visiting professorships at the University of Pittsburgh and in Norway, Austria, and Italy. His work has been translated into Italian, Chinese, Korean, Japanese, and English.

ERIC HAYOT is distinguished professor of comparative literature and Asian studies at the Pennsylvania State University and author of four books: *Chinese Dreams, The Hypothetical Mandarin, On Literary Worlds,* and *The Elements of Academic Style.*

LEA PAO is assistant professor of German at Stanford University.